Utopias and the Environment

Utopias and the Environment explores the way in which the kind of 'dreaming', or re-visioning, known as the 'utopian imaginary' takes environmental concerns into account. This kind of creative intervention is increasingly important in an era of ecological crisis, as we witness the failure of governments worldwide to significantly change industrial civilization from a path of 'business as usual'.

In this context, it is up to the artists – in this case authors – to imagine new ways of being that respond to this imperative and immediate global issue. Concurrently, it is also up to critics, readers, and thinkers everywhere to appraise these narratives of possibility for their complexities and internal conflicts, as well as for their promise, as we enter this new era of rapid change and adaptation. Because creative and critical thinkers must work together towards this goal, the idea of the critical utopia, coined by Tom Moylan in response to the fiction of the 1970s, is now ingrained in the common argot and is one of the key ideas discussed in this book. This development in the genre, which combines self-reflexivity and multiple perspectives within its dreaming, represents the postmodern spirit in its most regenerative aspect. This book is testament to such hopes and potential realities.

This book was originally published as a special issue of *Green Letters: Studies in Ecocriticism*.

Geoffrey Berry is an independent researcher, writer and public speaker promoting ecological identity in response to the anthropocene.

Utopias and the Environment

Edited by
Geoffrey Berry

Routledge
Taylor & Francis Group

LONDON AND NEW YORK

First published 2016
by Routledge
2 Park Square, Milton Park, Abingdon, Oxon, OX14 4RN, UK

and by Routledge
711 Third Avenue, New York, NY 10017, USA

First issued in paperback 2017

Routledge is an imprint of the Taylor & Francis Group, an informa business

British Library Cataloguing in Publication Data
A catalogue record for this book is available from the British Library

ISBN 13: 978-1-138-29485-1 (pbk)
ISBN 13: 978-1-138-93802-1 (hbk)

Typeset in Times New Roman
by RefineCatch Limited, Bungay, Suffolk

Publisher's Note
The publisher accepts responsibility for any inconsistencies that may have
arisen during the conversion of this book from journal articles to book chapters,
namely the possible inclusion of journal terminology.

Disclaimer
Every effort has been made to contact copyright holders for their permission to
reprint material in this book. The publishers would be grateful to hear from any
copyright holder who is not here acknowledged and will undertake to rectify
any errors or omissions in future editions of this book.

Contents

Citation Information

The following chapters were originally published in *Green Letters: Studies in Ecocriticism*, volume 17, issue 3 (November 2013). When citing this material, please use the original page numbering for each article, as follows:

Editorial
Guest Editor's introduction
Geoff Berry
Green Letters: Studies in Ecocriticism, volume 17, issue 3 (November 2013) pp. 195–199

Chapter 1
Utopias, miniature worlds and global networks in modern Scottish island poetry
Garry MacKenzie
Green Letters: Studies in Ecocriticism, volume 17, issue 3 (November 2013) pp. 200–210

Chapter 2
(Mis)Reading Hardy: 'Wessex' as green imaginary
Adrian Tait
Green Letters: Studies in Ecocriticism, volume 17, issue 3 (November 2013) pp. 211–222

Chapter 3
Utopian Zionist development in Theodor Herzl's Altneuland
Nicola Robinson
Green Letters: Studies in Ecocriticism, volume 17, issue 3 (November 2013) pp. 223–235

Chapter 4
'The Republic of Arborea': trees and the perfect society
Shelley Saguaro
Green Letters: Studies in Ecocriticism, volume 17, issue 3 (November 2013) pp. 236–250

Chapter 5
Hope of a hopeless world: eco-teleology in Margaret Atwood's Oryx and Crake *and*
The Year of the Flood
Nazry Bahrawi
Green Letters: Studies in Ecocriticism, volume 17, issue 3 (November 2013) pp. 251–263

Chapter 6

Genre, utopia, and ecological crisis: world-multiplication in Le Guin's fantasy
Katherine Buse
Green Letters: Studies in Ecocriticism, volume 17, issue 3 (November 2013) pp. 264–280

Chapter 8

Afterword: the utopian dreaming of modernity and its ecological cost
Geoff Berry
Green Letters: Studies in Ecocriticism, volume 17, issue 3 (November 2013) pp. 281–294

The following chapter was originally published in *Green Letters: Studies in Ecocriticism*, volume 5, issue 1 (January 2004). When citing this material, please use the original page numbering for each article, as follows:

Chapter 7

The Biologisation of Ecofeminism? On Science and Power in Marge Piercy's Woman on the Edge of Time
Martin Delveaux
Green Letters: Studies in Ecocriticism, volume 5, issue 1 (January 2004) pp. 23–29

For any permission-related enquiries please visit:
http://www.tandfonline.com/page/help/permissions

Notes on Contributors

Nazry Bahrawi is a faculty member of the Humanities, Arts and Social Sciences programme at the Singapore University of Technology and Design, Singapore, established in collaboration with the Massachusetts Institute of Technology, USA. He is also a research fellow at the Middle East Institute, National University of Singapore as well as a contributing editor of *Critical Muslim*, a quarterly journal of ideas and issues which presents Muslim perspectives on the great debates of the contemporary world.

Geoffrey Berry is an independent researcher, writer and public speaker promoting ecological identity in response to the anthropocene.

Katherine Buse is a PhD candidate in Literature at UC Davis, California, USA. As a Marshall scholar, she earned an MA in Science Fiction Studies from the University of Liverpool, UK, in 2012 and an MPhil in Criticism and Culture from the University of Cambridge, UK, in 2013. She works at the intersections of cultural studies, ecology, and speculative fiction.

Martin Delveaux is head of Modern Languages at Fettes College, Edinburgh, UK. He completed a PhD in English Literature, and a PGCE in Modern Languages, at the University of Exeter, UK.

Garry MacKenzie is a PhD candidate at the University of St Andrews, UK. His thesis examines the role of landscape in modern and contemporary poetry. His research interests include twentieth-century and contemporary poetry, Scottish literature, eco-criticism, and children's literature.

Nicola Robinson is a PhD candidate in the Department of English and Related Literature at the University of York, UK. She is working on a thesis entitled 'Resisting development: land and labour in Israeli, Palestinian and Sri Lankan literature'. She has also published an essay on Sri Lankan literature in the *South Asian Review*.

Shelley Saguaro is professor and head of Humanities at the University of Gloucestershire, Cheltenham, UK. She is the author of *Garden Plots: the politics and poetics of gardens* (2006) and has published articles and chapters on gardens, trees, and co-evolutionary poetics.

Adrian Tait completed his PhD in 2011, entitled 'An Eco-critical Approach to the Poetry of Thomas Hardy: from Wessex Poems to Time's Laughingstocks'. He continues to research Hardy's poetic vision of the environment, and he has published related papers in *Green Letters* (2011) and *The Hardy Society Journal* (2012).

INTRODUCTION

It has been widely acknowledged, since Ernst Bloch identified a utopian impulse in phenomena as widely varied as religious and philosophical tracts, daydreams and high art and advertisements, that Thomas More in 1516 named both a new genre and an ancient instinct. That a special issue of *Green Letters* should be dedicated to exploring the way the kind of dreaming known as the utopian imaginary intersects with environmental concerns should come as no surprise. This is especially so because, in an era of looming ecological crisis, the inability of governments worldwide to change business as usual stands as an enormous failure of resolve. It is up to the artists – in this case authors – to imagine forwards new ways of being that respond to an issue so great it potentially threatens to compromise the ability of life to flourish on this planet. And it is up to critics, scholars, writers and thinkers everywhere to consider these possibilities, their complexities and built-in conflicts, and the sociopolitical and environmental contexts within which such experiments may be supported or condemned. It is because creative and critical thinkers must work together on this issue that the idea of the critical utopia, coined by Tom Moylan in response to fiction of the 1970s, has become so ingrained in the common argot and emerges so often in this special issue. Here was a new development in the genre that combined self-reflexivity and multiple perspectives within its dreaming, representing the postmodern spirit in its most regenerative aspect.

It might be worth taking a moment to revisit the core elements of this invaluable intervention. This new and ongoing form maintains the traditional hope of utopia – that the creative and responsible human spirit could prevail over our inevitable flaws and destructive elements – while embracing many of the pivotal qualities of postmodernism. Critical utopias are open systems, self-aware, responsive to challenges to utopia but unwilling to give up on it, resisting anti-utopia and challenging 'any tendency toward a narrowly conceived and enforced utopianism' (Moylan 2000, 82–83). In the critical utopia, self-reflexivity, self-analysis, self-critique, realism about the flawed nature of any collective decision-making process (aka politics) and indeed of human nature in general, and relativity amongst alternative belief systems can all be understood as paths towards greater individual and collective awareness. As limited beings, in a situation of social interdependence with others, human individuals can, according to this hopeful genre, know themselves better, find a deep sense of satisfaction and develop qualities that bolster commitment to a civic sense of duty. Ruth Levitas followed Miguel Abensour to comment on the associated 'education of desire' at length and of course this discussion remains a vital part of the translation from a commitment to a broad sense of social justice to the ecological consciousness that may benefit from an ecotopian dreaming.

Moylan reiterated that 'awareness of the limitations of the Utopian tradition' means that 'these texts reject Utopia as a blueprint while preserving it as a dream' (1986, 10–11). 'Finally', Moylan concluded, such novels focus 'on the continuing presence of difference and imperfection within the Utopian society itself and thus render more recognizable and dynamic alternatives' (10–11). This leaps beyond the straw man argument used by those conservative thinkers opposed to the utopian imaginary on the grounds that utopias describe a kind of perfect existence (Sargent 1994, 5–6). Fictional worlds could also tie this new 'ideal' sense of community to the more-than-human world, often in ways that neatly avoid the romantic allure of the pastoral. Texts such as *Ecotopia* by Ernst Callenbach, Ursula K Le Guin's *The Dispossessed*, Marge Piercy's *Woman on the Edge of Time* and Kim Stanley Robinson's *Pacific Edge* all combined qualities of the utopian impetus with the kinds of realism about human nature Moylan could see emerging, to provide a richly imaginative counterweight to the dominant paradigm and its unsustainable, one-dimensional fantasy of mastery and profit.

Much has changed in the intervening decades, yet the general outlines of the argument remain fairly consistent. Fredric Jameson provided another major development when he moved beyond Karl Mannheim's dualism to recognise that there can be no simple split between the 'ideological', which serves to support the existing state of affairs, and the 'utopian' spirit of revolution. For Jameson, *any* political ideology draws from the wellsprings of the utopian impulse, including those defending hegemonic ruling class interests (Jameson 1981, 291). In this, he explicitly followed Bloch, who saw evidence of this ubiquity everywhere. So, the task that remains is to analyse any text, literary or otherwise, for the *kinds* of utopian impulses they carry or dismiss, especially when crisis calls for sustained action, as is the case in terms of climate change today. Class distinctions remain central to any substantial sociopolitical or economic critique, but they are not the only ones important to this discourse.

The articles presented here offer a variety of ways that texts, whether contemporary or of historical value, can be interpreted through the multiple lenses of utopian and dystopian theoretical positions. They represent a sophisticated exploration of what we may call the critical ecotopia, or more often the critical ecodystopia. The former of these has been defined by David Barnhill as requiring three qualities: it should be specifically positive about the natural world; it should be a positive social eutopia; and it should bring together human systems of life and the ecological systems they are a part of. This would include interactive harmony with ecological systems so that there is mutual flourishing (Barnhill 2011, 129). Barnhill goes on to clarify that such works should make a critique of the contemporary world as well as of the imperfection of 'a particular ecotopia or ecotopianism in general' (137). Obviously a similar parallel operates for the critical ecodystopia – it should offer some hope that ecological conditions today can be improved with reference to the horror that awaits us if the situation does not change for the better in the present (or near future).

These articles both satisfy and go beyond these necessary and sufficient conditions of an investigation into literatures of utopianism, dystopianism and the environment such as class, power relations and specifically ecological concerns. They further investigate ways that ideological assumptions can be projected onto idealised landscapes and rejected by those inhabiting the same places, while complicating the way associated genres such as the pastoral can still offer examples of important ecological concepts such as the bioregion. They investigate the way ancient agricultural biases lionising 'mastery' over the earth continue to be perpetuated in modern fiction and politics. They think further through the human relationship with non-human nature, the way our languages encapsulate and embody the power of definition and the intimate links between words, attitudes and behavioural patterns. Finally, they also consider the relationship between technology and faith. It is as if, then, we had drawn up a wish list of the kinds of issues felt most immediate to the way that the ecological humanities respond to the climate crisis and requested that they be dealt with in terms of the utopian imaginary. But of course these things are open to chance and I would like to thank the collection of authors who have submitted these articles for the broad and effective spread that their articles represent.

As expected, the articles themselves deal with the way we read utopian literature today, employing significant theorists when appropriate. It seemed a good idea, then, to add an Afterword that considered in more detail the way that utopian theory itself has and could respond to the ecological crisis on behalf of concerned citizens everywhere. Hopefully that piece can stand as an equivalent intervention in the light of these excellent articles.

Garry MacKenzie's piece on modern Scottish island literature shows how certain poets play with, investigate and subvert notions of the pastoral that can be too easily projected upon places of isolated natural beauty in an era of instability and climate change. Garry MacKenzie points out, for example, that the idea of the island of St Kilda as a space outside of modern society allows it to stand as 'an elsewhere from which to critique a century of western civilisation, and in particular contrasts its apparent ideals with the poet's contemporary British society'. The poems he discusses also relate the cruelty of life close to the elements, reminding us that the ancient art of place-naming is closely associated with prodigious knowledge of the land and its creatures, referring to 'hardship, ruin and weeping as well as to virtue and meetings between lovers'. The projection of human values onto place has rarely been better handled and Garry MacKenzie adds also to the discourse around the way local languages tussle with global concerns and touristic desires for an escapist paradise by the sea.

Following this, Adrian Tait investigates the way that Thomas Hardy's fictional 'Wessex' might be interpreted through the lens of bioregionalism. Adrian Tait asks us to reconsider the way Hardy's fictional place, which might easily be considered from within a rubric of exclusionary entity, also suggests a more fluid and adaptable example of 'ecological thought'. The scale at which Hardy considers the individual and their agency may, in fact, represent an aspect of ecotopian hope as well as compensatory pleasure to urbanised readers. Adrian Tait wants to acknowledge the extent to which Hardy instilled a 'disruptive surplus' into Wessex that would have made Bloch proud. This question inspires Adrian Tait to employ de Geus's definition of the ecological utopia, which requires a distinction to be drawn between utopias of sufficiency and utopias of abundance. The bioregionalism of Kirkpatrick Sale represents a well-known intervention around templates of bioregionalism as a potential utopia of sufficiency and Adrian Tait acknowledges Hardy's *The Woodlanders* as a fine example of the way people who live in close contact with a place are best positioned to describe its intricacies and the possibilities for bioregional 're-inhabitation'.

Next, Nicola Robinson considers the development of a utopian society in the light of Theodor Herzl's *Altneuland*, opening with a discussion of David Blumenthal's commentary on Salvador Dali's paintings about the settlement of Israel, depicted as a place of challenging difficulties and even 'retarded agricultural development'. Her analysis shows the way that a colonising project can be infused with a vision of settler-nationalism, in this case Jewish. What struck Robinson as noticeable was the fact that the same commentary was used for the restaging of this exhibition in 2011, indicating that 'this idea of Zionist development discourse and practice as motivated by ethno-nationalism is a recurring one and still being unproblematically accepted and disseminated today.' Herzl's novel of 1902, later published in English as Old-New-Land, provides a utopian bildungsroman that projects a satisfying vision onto an often difficult reality. Robinson concludes that, while Herzl paints a negative picture of the environment and original inhabitants of the land being settled, his utopian novel also shows that there can be a 'peaceful and mutually beneficial co-existence between Jews and Palestinians'.

Shelley Saguaro's piece shifts us from a focus on a harsh land seemingly in need of developmental improvement to an ancient locus of utopian longing, liberty and individual authenticity – the forest. Saguaro considers two texts, Italo Calvino's *The Baron in the Trees* (1957) and Sam Taylor's *The Republic of Trees* (2005). She points out the way that both feature youthful rebels seeking freedom from the almost unstoppable impact of a new, profit-driven machine age on treed landscapes and a sense of identity alike. Fired with Rousseauian revolutionary enthusiasm, both protagonists choose life amongst the trees rather than the inevitable and sometimes unacceptable compromises of society. With reference to a wide variety of relevant texts, Saguaro takes us on a journey through the forest of these authors' imaginations. It is not always comfortable travelling, nor should it be, but the concluding section branches out into the Ecuadorian Declaration of the Rights of Nature announced in 2008 with the aim of it being incorporated into the national Constitution. Alberto Acosta, Chair of the Assembly at the time, minister of energy and a presidential candidate in 2013, actually referenced Calvino in his attempt to shift the national consciousness from anthropocentrism to a biocentrism. In this he also aimed towards a deconstruction of Eurocentric ideologies that privilege humanity's domination over the rest of nature.

Moving more concretely into the realm of science fiction, Nazry Bahrawi reads Margaret Atwood's fabulously successful novels *Oryx and Crake* and *The Year of the Flood* for the element of critical dystopia they convey through their otherwise bleak exploration of the near future. Bahrawi maintains that the utopian lining of this dark cloud takes the form of an eco-teleology, which critiques and responds to the 'scientism' that assures our collective destruction at the hand of Crake and his substantial funding bodies within the gated compounds. Bahrawi uses Patrick Curry as a guide in critiquing this familiar version of human mastery and sees a form of Darwinian natural selection directing the way the God's Gardeners blend religion with biology. Bahrawi points out that the Crakers' spontaneous creation of symbolic narrative is a reminder of the immutability of story even in 'a wild posthuman ecology'. In spite of the devastated landscape of the post-apocalyptic sections of these novels, Bahrawi notes that the Blochian idea of the 'Not-Yet' retains its utopian promise, never truly free of religious impulse or evolutionary pressures, but embodied as hope in a hopeless world.

Adding fantasy and an activist classic to our consideration of science fiction, Katherine Buse offers an inventive reading of the novels of Ursula K Le Guin alongside the environmental non-fiction (and occasional creative flourishes) of Rachel Carson's work of inestimable environmental value, *Silent Spring*. The focus on Le Guin centres around *The Farthest Shore*, one of the texts that made up her very popular *Earthsea Quartet*. Buse outlines the way that, for Earthsea wizards, magical language arises out of a close attention to the life of the land, revealing both a mystical depth and an appreciative, unflinching gaze into its intricacies, conflicts and complexities. Not only this, but a thing's 'true name' speaks of its intrinsic worth, beyond its use in human hands. Another aspect of Buse's discussion that will be of interest to those interested in utopianism and ecology lies in her defence of fantasy, as well as more serious science fiction, as a substantial vehicle for such ideas in the current era, especially in reference to the critical work of Darko Suvin. Perhaps surprisingly, Buse wants us to see the mythic element in Carson's work, as well as the better known anthropological foundations of Le Guin's. Just as Carson, Le Guin and the characters in their texts do, we need to remain awake to the signs of ecological crisis, with every aspect of our intelligence we can muster, if we are to continue to offer alternatives to the dominant paradigm of production and consumption, which is steeped deeply within the profit motive of mastery over non-human nature.

If this collection helps readers to remember that the utopian spirit arises spontaneously within each of us, it may also help to remind us that this can be put into the service of complacency as well as activism. Utopian imaginaries abound in contemporary culture, but those that carry the weight of the capitalist agenda are granted far more 'realism', despite their unsustainability, than those that have thought through the problematics of human societies in relation to the rest of nature and responded with a critically self-reflexive set of possibilities. My Afterword takes this issue up with specific reference to utopian theory (especially Blochian), my own work on light as a symbol of prestige, progress and faith, and the possibility that alternative epistemologies can contribute to the debate around the value of ecotopian imaginaries in an age of ecological crisis. Much of the research for this work was undertaken during my time as Erasmus Mundus Post-doctoral researcher at the Ralahine Centre for Utopian Studies, based at the University of Limerick in Ireland. Many thanks to Tom Moylan and the rest of the scholars there for their time, expertise and generosity of spirit, as well as to visitors including Lyman Tower Sargent and Ruth Levitas. Parts of my Afterword were presented at the Composting Culture Conference of ASLE-UKI (the Association for the Study of Literature and Environment (UK and Ireland)) held at the University of Worcester in September 2012. I must finally express my gratitude to the funding body EUOSSIC (the EU-Oceania Social Science Inter-regional Consortium) for the opportunity to undertake this fascinating and hopefully valuable research.

References

Barnhill, D. 2011. "Conceiving Ecotopia." *Journal for the Study of Religion, Nature and Culture* 5 (2):126–144.
Jameson, F. 1981. *The Political Unconscious; Narrative as a Socially Symbolic Act*. London: Methuen.
Moylan, T. 1986. *Demand the Impossible: Science Fiction and the Utopian Imagination*. New York: Methuen.
Moylan, T. 2000. *Scraps of the Untainted Sky*. Boulder, CO: Westview Press.
Sargent, L. T. 1994. "The Three Faces of Utopianism Revisited." *Utopian Studies* 5 (1):1–37.

Geoff Berry
The Phoenix Institute of Australia, November 2013

Utopias, miniature worlds and global networks in modern Scottish island poetry

Garry MacKenzie

School of English, University of St Andrews, St Andrews, Fife, UK

This article considers the extent to which modern Scottish island poetry constructs a utopian landscape. The relative isolation of islands makes them suited to utopian narratives: poetry about the remote St Kilda can both highlight and subvert its utopian potential as a model of a pre-modern ecological society. I discuss how these themes emerge in St Kilda poems by Douglas Dunn and Robin Robertson. I then situate the contemporary Shetland of Jen Hadfield's poetry in relation to the notion that Scottish islands retain elements of a pastoral 'Golden Age', as is suggested by the tourism industry. Finally, I discuss how debates within ecocriticism about the significance of place and the interconnections facilitated by globalisation can be informed by Robert Alan Jamieson's non-insular Shetlandic poetry, which is written in local dialect but retains an unapologetically global perspective.

The literary utopia, argues Tom Moylan in *Demand the Impossible*, is a form which is particularly suited to periods of rapid social change and uncertainty. It enables the reader to consider 'what is and what is not yet achieved', by envisioning an alternative society distant from the author's own in time or space (Moylan 1986, 3). Moylan suggests that such utopias may be other worlds, past or future societies or, as in the case of Thomas More's Utopia, remote islands (3). The locations Moylan proposes are all closed spaces, impermeable to external influences which would potentially dilute their difference, and so these locations remain 'elsewhere', if not quite the 'nowhere' that the name of More's fictional island implies. In the globalised and postmodern contemporary world which, as Fredric Jameson points out, it is difficult for any individual to comprehend in its totality and locate their position within (1991, 34–35), the appeal of these bounded, easily-defined, alternative territories may be particularly strong. Given the complexity of factors affecting climate change, the utopian appeal of a small and clearly defined place may increase even further if its society can suggest a more environmentally friendly model for interactions between humans and their environment. However, as predictions of the effects of climate change make the future appear increasingly bleak, the temptation is to locate this utopia in the pastoral past.

Islands lend themselves to utopian narratives because they can be portrayed as miniature versions of the mainland or, indeed, of the world. In *On Longing*, her work on the aesthetics of the miniature and the gigantic, Susan Stewart writes that any miniature has to be an island, in that it is cut off by sealed borders (1993, 68). This creates a manageable, diminutive version of experience; we can know all of a miniature object, whereas we can only ever ascertain part of something gigantic (71). The small object increases in symbolic significance because we can observe the spatial closure of a perfect form, a particularised object that comes to stand for other instances of its occurrence. Hence, the miniature has the capacity to inspire reverie, transporting the observer to an apparently permanent world (47–48). Whereas Stewart uses the island as a metaphor for enclosure (an idea which is not wholly accurate given that islands are not impermeable), the concept of the miniature can be applied to islands themselves; in the limited space of the island, details acquire more significance. This is not just an aesthetic argument – as biogeographers have shown, islands offer simplified models of the world's ecological complexity, enabling study of population dynamics, migration and extinction and the development of theories that can be applied to continental landmasses where species populations are increasingly fragmented by human development (Whittaker 1998, 1,4).

There is more than a hint of both utopian and pastoral imagery associated with the rural and generally sparsely populated islands to the west and north of Scotland. Whilst I acknowledge the complexity of the pastoral tradition, I use the term 'pastoral' here to mean both the portrayal of the rural as a tonic to urban modernity and the oversimplification of rural life in ways that overlook environmental and social conflicts. Their separation from the British mainland means that Scottish islands are physically distant from metropolitan centres of power and influence, and this is part of their appeal; they offer an iconography that is at least slightly different from that of the rest of the United Kingdom. Lewis-born poet Kevin MacNeil, in his introduction to a recent anthology of Scottish island poetry, contends that this 'sense of being "other" [...], removed from the heart of things' enables the islander to become an observer of mainstream British culture, able to offer critique and comment from the periphery 'where opposites clash or converge, where creativity and danger are at their most alive' (2011, xxi). Both islands and their people are, therefore, given utopian potential by MacNeil. While his point is primarily about legitimising alternative cultural and linguistic traditions that are marginal to (but hardly completely separate from) dominant British cultural discourse, he makes this point by appealing to the image of Scottish islands that is most potent within that discourse. The phrase 'clash and converge' carries the suggestion of the sublime Hebrides of the Romantic imagination, with crashing waves and towering cliffs. He also overplays marginality, since cultural differences and physical distance have been made less significant by modern communications and media as well as by migration. The problem of perspective, as so often in writing about rural Britain, emerges here – MacNeil's argument both critiques and reinforces the notion of island peripherality.

For the visitor, the Scottish islands offer such abstractions as isolation, a slower pace of life and beautiful scenery: the website of VisitScotland, the national tourist board, promises that 'you can experience the tranquillity of island life' in 'nature's paradise'. This lure has been potent for a long time. As early as 1730, in the *Autumn* section of *The Seasons*, James Thomson praises the 'naked melancholy Isles' on which the 'harmless Native [tends] his small Flock'. This 'Shepherd's sea-girt Reign' is remote, sublime and (despite the typically bad weather) characterised by the simple pastoral life of its inhabitants (1981, 178–179). This perspective has long been problematic: in 1773, Boswell and Johnson toured the Scottish Highlands and Islands and had their expectations challenged by the social and environmental upheaval, brought about by changes in rural land use, that they witnessed. Johnson's record of the trip meditates upon the material and intellectual needs of the Highland peasantry rather than celebrating their pastoral simplicity (1984, 102–103). Nevertheless, in the nineteenth century, with tourists flocking to the Highlands and Islands to experience for themselves the landscape of Romantic fiction and poetry, the first cruise ships sailed to the most distant of the Outer Hebrides, St Kilda, seeking a glimpse of the life outlined by poets like Thomson.

For those whose everyday concerns are located in this landscape, talk of tranquillity and paradise can mask social and environmental concerns. In 1939, Shetland-based poet Hugh MacDiarmid bombastically declared that he was 'all for the de-Thibetanisation [*sic*] of the Scottish Highlands and Islands', meaning that these regions should not merely be preserved as retreats in which to cultivate an enlightened state of mind (18). He also argued that greater autonomy would allow the outlying islands of Britain to pursue development that was alternative to British political norms and so bring about a radical, even utopian, change (6). In the 1980s, Iain Crichton Smith, another Lewis-born poet, used his essay 'Real People in a Real Place' to condemn what he saw as the falsehood of the tranquil Hebridean paradise, home to Gaelic-speaking people patronised in films such as *Whisky Galore* as either 'noble savage[s]' or like 'little children' unsuited to the demands of the modern, urban 'real world' (1986, 14–15). Such patronage, he says, relies on the assumption that these 'natives' have never actively chosen to live where they do – otherwise they would have relocated to the 'real world'. Instead, they 'must be considered as having always lived in this enchanted unreal world, with its mountains and its lochs and its sunsets' (15). Smith argues that the portrayal of the Hebrides in tourist literature and in the songs and stories of émigrés obscures the 'difficult' living conditions of a place which 'may have been a home but was never an Eden' (18). He contests the construction of the sentimental Scottish island in his poetry as well as his prose. For example, 'Poem of Lewis' embraces what Terry Gifford has called anti-pastoral elements: bleak island land-scapes in this 'black north' are home to equally bleak Calvinist dogma by which poetry, compassion and gaiety are 'like a shot bird' unable to take flight (Smith 2011, 3). In another poem simply called 'Lewis', the 'eternal whine' of the wind is a dominant and recurring image, heard across an island where exhausted elderly people work the land in bad weather (256). To quote Gifford's summary of Matthew Arnold's anti-pastoralism, 'the natural world can no longer be constructed as "a land of dreams"' (1999, 120). Smith argues that Scottish islands should not be seen as pastoral utopias. As I will discuss below, contemporary poets such as Jen Hadfield and Robert Alan Jamieson continue to negotiate more nuanced perspectives on island life and landscape in their work.

As these comments show, there is a conflict established in writing about Scottish islands between an idealised construction of social and ecological relations and reactions against it. This may be the case in all writing about rural Britain as it wrestles with its pastoral heritage. Island writing is different, however, because its subject lends itself as a microcosm for wider environmental and social concerns – whether these are utopian aspirations, biogeographical research or arguments about pastoral constructs. At the same time, the Scottish islands should not be seen as isolated 'elsewheres', as they remain part of the British archipelago and retain cultural connections with places across the United Kingdom and the world. Much of the poetry from and about the islands offers close observation of human and non-human ecology, but retains a resolutely outward-looking gaze. This simultaneous separation and interconnection makes them an appropriate locus for consideration in debates about whether environmental concerns should best be understood through proximity to place or through embracing global networks of communication and ideas. I will look first at the utopian potential of islands in poetry about St Kilda, before turning to work that embraces both the local and the global.

Utopian St Kilda

In some ways, the archipelago of St Kilda is the exemplar of the Scottish island group: difficult to reach, dramatically beautiful, endowed with abundant birdlife and rare sub-species and culturally distinct from metropolitan Britain. Now a dual UNESCO World Heritage Site recognised for both its ecological and cultural significance, St Kilda is a tourist destination of international importance. Like many other Scottish islands, it was an enclave for Gaelic culture and language marginalised from the British, and even Scottish, mainstream, but its small population, squeezed by emigration and changes detrimental to the sustainability of life there, requested to be evacuated and were entirely relocated to the mainland in 1930. St Kilda retains a hold on the popular imagination: books such as Tom Steel's *The Life and Death of St Kilda*, Charles Maclean's *Island on the Edge of the World* and Donald Gillies' and John Randall's *The Truth about St Kilda* have speculated on the decline of this island society and have been frequently reprinted.

Unsurprisingly, St Kilda is also an appealing subject for modern Scottish poetry, which either embraces or revises its romantic allure. In his poem 'St Kilda's Parliament: 1879–1979', first published in 1980, Douglas Dunn imagines a photographer looking back at his 1879 picture of a gathering of islanders with the benefit of a century of hindsight (2003, 69–71). His subjects do not live a life of pastoral tranquillity – their lifestyles are determined by the environment in which they live, the 'roaring gales', the 'diet of solan goose and eggs', and 'their companionship with rock' (69). These factors, the photographer speculates, inform the expressions on the men's faces as they watch the cameraman. Despite their shyness, and physical similarities arising from the small gene pool, the observer comments that 'each is individual': his seeming intention is to not cast each islander as a 'type' or to come to any firm conclusions about whether these are 'Wise men or simpletons' (69). In other words, his description attempts to avoid the elision of difference between rural people which Raymond Williams sees as a hallmark of pastoral poetry (1973, 257), and which is condemned in Smith's call for literary attention to be given to the Hebrides' 'real people'.

In Dunn's poem there is, however, a tension between this aim and its success: the islanders are still repeatedly referred to as a group, as 'they' and 'them'. Verbs such as 'look' and 'stare' dominate over more intimate forms of connection. None of the islanders speak in the poem, but watch the camera silently 'like everybody's ancestors', thus becoming an image of archetypal humans, representatives of the race (70). The speaker is not one of the St Kildans, nor can he speak for them: he admits that he cannot understand their Gaelic speech and can only speculate as to what their expressions mean. Their intimate knowledge of their islands' ecology, from the huge gannetries the community relied upon for food to the indigenous sub-species of the tiny 'St Kilda mouse and St Kilda wren', is doomed to 'fall into the texts of specialists' (71). There is a great deal of distancing going on in the poem, and this effect is furthered by the overall conceit: the speaker is looking at a photograph, not the place itself, and is looking from a century away.

Such distance befits the elegiac tone of a poem about an isolated community which has ceased to exist except in old photographs. Distance, however, also gives this poem its utopian feel. Although poverty is shown to afflict the islanders, Dunn still depicts a society uncorrupted by class inequalities that are reinforced by indicators of education and refinement, by 'Hierarchies of cuisine and literacy' (70). Instead of economic growth, the speaker argues that these late nineteenth-century people desire only to maintain their 'eternal/Casual husbandry', their long-established subsistence economy developed in sympathy with their fragile and demanding environment (70). The speaker juxtaposes this minute island civilisation with atrocities elsewhere, from the Anglo-Zulu War of 1879 to the destruction of European cities in twentieth-century wars (71). The implication is that St Kilda offered an alternative, more egalitarian, environmentally sustainable model for society, or at least represents the potential for that society to be realised. The title of the poem, referring to the daily gathering of workers to decide the day's work, also suggests devolved regional government and alludes to the Scottish devolution referendum of 1979. The speaker in the poem uses St Kilda as an elsewhere from which to critique a century of western civilisation, and in particular contrasts its apparent ideals with the poet's contemporary British society.

Other more recent poetic accounts of St Kilda are less utopian in outlook. As Edna Longley (2010, 151) notes, Don Paterson's 'St Bride's: Sea-Mail' can be read as a 'dystopian riposte' to Dunn's St Kilda, in which islanders over-exploit their environment by using bird carcasses for sport until the population on which they depend is wiped out (Paterson 2003, 2–3). In a similar vein, Robin Robertson's 'Law of the Island' describes the sadistic punishment enacted by another St Kilda-like community. A man is bound and floated in the sea, weighted down so that all but his head is submerged. Live mackerel are tied in front of his eyes, and the islanders wait on the shore for a gannet to dive from high above and spear the fish (2010, 33). An admittedly eco-friendly death penalty, this image subverts any idea that islanders live in a paradisal utopia.

It is, however, another of Robertson's poems that offers the least fanciful vision of St Kilda, one which suggests a human ecology based on familiarity and intense locality. In 'Leaving St Kilda', the speaker is watching the archipelago as he departs in the 1930 evacuation (2010, 25–29). There is no history or analysis in the poem, except what is contained in the place names and physical description of the island, which are both incredibly rich. Every cleft and rise in the landscape seems to be named, either in Gaelic, English or both, and these names signify something of the natural and human history of the place. In a moving litany, the speaker names many such landmarks, including 'Stack of the Guillemot', 'Cleft of the Seals' and 'Conachair the Roarer' (25). Another place, 'Ruaival', is translated as 'the Red Fell, pink with thrift', the coastal wildflower whose name may also be alluding here to the necessary frugality of life in St Kilda (27). An almost erotic familiarity with his environment is demonstrated in the speaker's naming, a sensation enhanced by the fact that several geographical features are named after parts of the body – the 'Mouth of the Cleft', 'The Heel', the 'Dale of the Breast' (26, 27). Many of the names convey information about which animals, birds and plants, relied upon for the community's survival, can be found where, with little distinction made between the St Kildan ecosystem and its society. Bird behaviour is often noted in tender asides: 'fulmars nest in their sorrel and chickweed' (25) – in this phrase, the possessive 'their', unnecessary for the literal sense, adds to its intimate tone. Later, the rock of Stac Lee is a 'stone hive of gannets, thrumming and ticking'. These gannets 'track the shadows/of the mackerel or the herring shoal' before diving in their hundreds (28). As well as paying attention to the biodiversity of the archipelago – four mammal, three plant and fifteen bird species are listed – human stories are also recalled as the speaker names such places as the 'Brae of the Weepings', 'The Milking Stone' and 'Skerry of the Son of the King of Norway' (25, 26). Myth and natural history blur in this litany: on 'Stac an Armin', the speaker tells us, the archipelago's last great auk was 'killed as a witch' (27).

The volume of detail in the poem, which for most of its four and a half pages reads like a list punctuated with digressions, contributes to its success in creating an oral cartography of St Kilda – in this project, it resembles such prose work as Tim

Robinson's studies of Aran and Connemara. This amount of detail means that the place cannot be easily dismissed as a sublime but empty wasteland now that it is depopulated. Nor can it be sentimentalised, as the account of the great auk shows. Place names refer to hardship, ruin and weeping as well as to virtue and meetings between lovers. Nevertheless, the attachment of St Kildans to their islands, and a sense of profound respect for their environment, is generated by the speaker's methodical rendering of each landmark; the tragedy of leaving St Kilda in this poem is the loss of this relationship between people and place.

The photographer in Dunn's poem ultimately remains distant from the community and environment which he records. Robertson's speaker achieves greater intimacy with the landscape and ecology of the islands. However, the history contained in the many place names does not alter the fact that the people must ultimately leave their island home, and it could be argued that naming cannot bring the speaker any closer to the non-human world he describes. The inclusion of Gaelic words unfamiliar to most readers within a primarily English text serves to emphasise the disjunction. 'Leaving St Kilda' nevertheless achieves the converse of this – the persistent naming and layers of human history that it contains bring the speaker, and perhaps the reader, closer to that environment. This is not an island utopia to aspire to in uncertain times, nor to nostalgically ascribe with merits lacked by our present society. Despite the wistful intimacy of the poem's speaker, an innate love of place and sense of primitivist 'dwelling' should not necessarily be ascribed to all St Kildans. Robertson's St Kilda can, however, be termed a utopia in the sense that it provides an alternative social model by which to assess the character and extent of our own relationship with our environment, in Moylan's phrase to consider 'what is and what is not yet achieved', and indeed what it is desirable to achieve.

Locality and internationalism

Robertson's St Kilda is home to a society that now exists only in the past. His poem contains little of the modern technological world. This is a trend seen in much Scottish island poetry and can lead to the assumption that any ecological message contained in poetry about rural Scottish islands will emphasise former ways of life that existed prior to the intrusion of machines, mass production and modern communication technology. Edwin Muir's 'The Northern Isles', for example, recalls an Orkney bathed in summer sunshine where agricultural labour is notably Arcadian, consisting of verdant pastures and abundant livestock ready for slaughter (MacNeil 2011, 3). Alastair Reid's 'Isle of Arran' reminisces about a childhood excursion in which reality dissolves into the world of daydream. The speaker is at the centre of a tranquil scene where bees sleep on the 'quiet quilt of heather', a butterfly flutters 'drunkenly' and the only discernible wind is the speaker's breath (2008, 32). George Mackay Brown's poetry, largely about pre-modern Orkney, exposes itself to the critique that it evades twentieth-century social and environmental realities.

This is not the only approach to Scottish islands found in poetry, however. In her 2008 collection *Nigh-No-Place*, Jen Hadfield revises the familiar tropes of island pastoralism with references to everyday modernity, examining and critiquing romantic archetypes with humour and with fidelity to her experience of place. An immigrant to Shetland (she was born in Cheshire), her poetry is inspired by Shetland's landscape, ecology and community, constituting what she refers to in her blog as an 'idiomatic mythology of the here-and-now'. An indication of this can be seen in her attitude to island work, which is neither romanticised nor sentimentalised. Hadfield has worked in the Shetland fish industry and responds to her experience in a series of 'Ten-minute break haiku'. Vivid details of gutting knives 'prattling' over cartilage, plucking 'Gut-worms' and holding a fish's 'quilted sphincter', which the poem's title suggests are recorded in brief interludes from work, are far removed from, say, Hugh MacDiarmid's aloof meditations about beauty and the 'primitive minds' of Shetland fishermen in his own poem 'Deep Sea-Fishing' (Hadfield 2008, 43; MacDiarmid 1993, 438). Hadfield is a poet unafraid to highlight the unpleasant and banal aspects of island life.

Such details do not dampen Hadfield's enthusiasm for Shetland. Her response to everyday island scenes is characterised in her poem 'Our Lady of Isbister' as a feeling of 'bliss' (46). Nevertheless, her Shetland is not somewhere to retreat to in order to gain enlightenment; neither is it a utopian alternative to mainland Britain. She rarely allows transcendent imagery to remain uncontaminated by the realities of modern island society. For instance, the poem 'The Wren' appears at times to envisage a mystical island where an 'Iron-Age sun' still shines and augury is possible: 'the ducks will gleam like curling stones. You'll get off scot-free', remarks the speaker. However, the landscape also contains a modern caravan, a polytunnel in which cabbages are grown for sale and a helicopter whose rotors echo in the opening and closing stanzas (47). In another poem, 'Hedgehog, Hamnavoe', the speaker meditates upon the creature quivering in her hands, which she examines like 'a crystal ball'. Where the reader may expect some kind of revelation, however, the speaker can only express determination that this encounter should have greater meaning than 'guesswork//and fleas' (48). In the process of offering various definitions for its title, the poem 'Gish' moves from the image of 'a channel of water strained through the wet red grass of a Fair Isle field' to 'a pish in the dark' (36). Modern technology and earthy humour repeatedly intrude to modify the reader's expectations of revelation or exultation in the landscape.

This process can also be seen in Hadfield's first collection, *Almanacs* (2005), which shares with *Nigh-No-Place* a close interest in island lives and landscapes. Images collide the 'natural' subject matter of Scottish island poetry with modernity, refreshing the reader's ways of seeing these rural, coastal places. For example, 'A970', a prose poem named after the main north–south road on Shetland's Mainland, uses horse metaphors to describe various vehicles seen on the road, from 'coltish GTIs' to the 'Clydesdale tyres' of hatchbacks (Clydesdales are a breed of horse traditionally used for pulling heavy loads because of their large size and strength). Whilst punning on the term 'horsepower', these images also associate modern transport with the older, more firmly pastoral means of travelling and of doing heavy agricultural work. This connection is made explicit in the closing sentence of the poem, 'The miles mount up like lazybeds.' Lazy beds are a form of cultivation involving parallel ridges and furrows. Once common in rural Scotland but now largely obsolete, their remains can still be seen at roadsides. Hadfield uses them here as an indication of the distance travelled by a speeding 'Girl Racer' on the A970, fusing Shetland heritage with everyday contemporary life (31).

In her strongest poems, Hadfield combines eloquent description of an island landscape with what Terry Gifford describes as post-pastoral tropes: a sense of awe of the natural world, awareness of the complexity of ecological processes and of the ways in which the human imagination is influenced by our environment, as well as a refusal to ignore technological modernity (Gifford 1999, 150–165). Another poem from *Nigh-No-Place*, 'Daed-traa', whose title derives from the Shetlandic term for 'the slack of the tide', is a close, imaginative observation of a miniature but vibrant world within the island's bounds, a rock pool. The pool contains a complex ecology: systems for regulating nutrients ('its ventricles') and miniature dramas of predation and survival ('its theatre' and its 'Little

Shop of Horrors'). There is an acknowledgement that the natural world is observed through the frame of culture as the crustaceans, plants and other creatures of the pool are figured in cultural terms, from the 'cross-eyed beetling Lear' to the 'billowing Monroe'. At the same time, the process of observation leads to another cultural product, the poem, as the speaker explains that she finds inspiration here. The pool may look still at the slack of tide, but is teeming with life and the variety of its contents 'mind me what my poetry's for', which in this case is a celebration of its biodiversity. Its 'crossed and dotted monsters' are both patterned predators and reminders of the process of writing, alluding to the idiom 'crossing i's and dotting t's', the meticulous completion of a document or task (2008, 35). This poem focuses in on a miniature ecosystem, a pool on an island shore, charging it with striking imagery but nevertheless through its diverse references expressing a complex and non-parochial relationship between people and place.

Instead of pastoral fulfilment on an island retreat, Hadfield writes in *Almanacs* and *Nigh-No-Place* about the recognition of a relationship between the inner life of the observer and external nature. The cosmopolitan frame of reference erases any notion the reader may have that her poetic Shetland is stuck in the past, or is a utopian society cut off from technology, communications, industry and popular culture that exist elsewhere. Her work is confidently post-pastoral and concerned with 'real people in a real place', to return to Iain Crichton Smith's phrase. It demonstrates the direction which the best Scottish island poetry of the future may take if it is to avoid an escapist depiction of landscape and environment.

Despite Hadfield's cosmopolitanism, a reading of the Scottish island poetry discussed above may lead to the conclusion that environmental imagination must be primarily predicated upon a close attachment to a particular locality, something that the island, because of its size and relative isolation, can offer. This argument sits well with the connection of utopias to specific places (Moylan, 32) and with such environmental ideas as bioregionalism, phenomenology and localist critiques of global capitalism. However, as Ursula Heise has shown in *Sense of Place and Sense of Planet*, the idea that environmentalism must rely on this 'ethic of proximity' is problematic (2008, 46). Not only does this approach fail to offer direction on issues that require global consensus and action but, she argues, it is also at odds with the shifting sense of place that individuals feel in our globalised, postmodern world (50–51). Instead, Heise proposes that environmentalism should move towards what she terms 'eco-cosmopolitanism', meaning that environmental discourse should embrace forms of belonging and activism that are relevant in today's interconnected world (59). Although she does not advocate a specifically utopian imaginary, her cosmopolitan vision is one which she argues society must aspire to achieve, and according to Heise, literature and criticism can suggest opportunities for a greater sense of 'eco-cosmopolitanism' (90). She is not alone in highlighting the problems of scale in environmental thought: thinkers such as Timothy Clark (2012) have sought to address the difficulty of relating local or individual actions to necessarily worldwide perspectives on environmental issues.

Island writing is well situated to respond to calls for a wider sense of global identity to balance the already prominent calls for a deeper sense of place. This is because, as Gillian Beer has shown, an island can be seen both as 'a final fixed object' – a unified and isolated space that might be utopian or dystopian – and as one of many points in 'large and complex systems of the present and the past' (2003, 32). The former paradigm lends itself to the idea of the island as a microcosm; the latter acknowledges that islands have long been 'crossroads, markets for exchange' and essential stopping points for sailors (33). Today, islands are more readily linked to other landmasses by boat and plane than ever before. Further, as biogeographers have shown, many species not only migrate to or from islands but can become established in previously unfamiliar habitats, leading either to an increase in biodiversity or to a catastrophic collapse of indigenous species (Quammen 1996). Interconnections exist at this ecological level and also at the social; the globalisation scholar Arjun Appadurai has argued that places should now be seen as processes within 'a world of flows' rather than as stable and coherent wholes (2001, 5–7). Rather than portraying islands as isolated, alternative communities and ecosystems, they should be seen as footholds in the global, bases from which to engage in the international transaction of ideas between cultures and languages and from which both local and cosmopolitan conceptions of environment can emerge. This is not to say that all aspects of globalisation should be embraced – the imperatives of global capitalism include patterns of production and consumption that have proved disastrous for ecosystems and their human inhabitants. However, a utopian vision for today should involve the repositioning of global systems of trade and ideas through such a cosmopolitan discourse, rather than imagining a discrete, ideal retreat.

This perspective can be seen to an extent in the scope of Hadfield's references, which range from the minute creatures of a rock pool to global popular culture. Island cosmopolitanism is more specifically addressed, though, in poetry by Robert Alan Jamieson, who writes Shetland-dialect lyrics which celebrate the internationalism of his native archipelago with a vocabulary and grammar unique to the village he grew up in. Each poem in his collection *Nort Atlantik Drift* is faced by a loose English translation and a prose introduction that sets its context; an audio-visual recording of his reading the work was also released to accompany the book. Beyond questions of accessibility, there is a dialogic element to this presentation, combining fidelity to a specific island location and its idiolect with a global language and perspective. Poems from *Nort Atlantik Drift*, Jamieson points out, are well travelled too, having been read in Shetlandic with translations to audiences in Bilbao, Tel Aviv, Tallinn and Prague (2007, 11–12). The content of the poems supports this project. 'Ta Kompis' ('To Compass') begins:

> Dæ kerriet da surfæs roondnis,
> da brædth a'da wirld,
> ati'da globbs a'dir heds. (78)

The single-line English translation of this stanza reads: 'They carried the full surface roundness, the breadth of the world, in the globes of their heads' (79). The 'roondnis' of the earth is paralleled with the 'globbs' of the sailor's heads; their minds contain the knowledge required to navigate the oceans but also an awareness of the variety, or 'brædth', of its places and cultures. Later in the poem, one sailor shows his family photographs he took at the Tokyo Olympics of 1964, while another describes crossing the International Date Line. To compass, as Jamieson argues in his introduction to the poem, is 'to measure, to make a circuit of' the world, as the Shetland seafarers of 'Ta Kompis' do while sailing to all points of the compass. The poet also stresses in his introduction that this island community living on the edge of Europe has been a cosmopolitan stopping point in global movements (77). Similarly, in Jamieson's 'Sievægin' ('Sea-Faring'), a Shetlander demonstrates 'a globbil awaarnis' ('a global awareness') because for him the ocean offers the limitless possibility of travel, the means to go 'quhaar du will' ('where you will'). The poem also draws on the image of the world as an island, the 'bloo baa' ('blue ball') first seen in photographs taken from space

(112–113). For the seafarer looking out across the ocean, this is not a new concept, but something he has already learned from a life of sailing the world.

The poems in *Nort Atlantik Drift* recall the lives of merchant seamen Jamieson knew as a child in the 1960s and attempt to preserve the dialect spoken by them and by the poet. They provide closely observed accounts of the relationships of people with the landscape in which they live – with its demands, dangers and kindnesses. It is not sentimentalised. Drowning is a constant danger for those working at sea, as the poem 'Ho's Bytirs' ('Shark's Teeth') makes clear (57–58); 'Botabit', which means the sharing of a boat's catch with the elderly or infirm, tells of the necessity of this custom because so many people 'hedna mukkil' ('haven't much') to survive on (44–45). Jamieson writes about a coherent community with a clearly defined way of living in their environment: in this, it resembles Robertson's St Kilda. What nostalgia there is in Jamieson's poetry is tempered by the lack of pastoral abstraction or generalisation, and his Shetland is no idyll. Neither is it cut off from the wider world. While Jamieson's poetry does not aim to provide an environmentalist message, it combines a close association to place with a cosmopolitan perspective, a dialogue between the local and the global which environmental literature and criticism is beginning to embrace more fully.

The Scottish islands remain contested landscapes, claimed as paradises in the hyperreal discourse of the tourism industry, in language that poetry often challenges and undermines. Hadfield's *Almanacs* and *Nigh-No-Place* deflate pastoral expectations and repeatedly remind the reader of the mediating role of culture in her understanding of Shetland's landscapes and biodiversity. On the other hand, both Dunn and Robertson, to varying degrees, utilise the utopian potential of the now lost human ecology of St Kilda in their poems about this archipelago. Dunn's poem emphasises St Kilda's subsistence economy and seemingly egali-tarian 'Parliament', although such a society remains in the distant, elegiac past. For Robertson, what remains of St Kilda after evacuation is a rich, sometimes vulnerable ecology, and a similarly rich and vulnerable body of memory contained in place names and anecdotes. Ultimately, however, neither poet views St Kilda as 'nature's paradise'; the notion of the idealising outsider is confronted by the role of Dunn's narrator and by the potentially unsustainable relation to place in Robertson's poem. As the poetry discussed above shows, the Scottish islands today are not utopias which offer greener, more holistic or pre-modern alternatives to our contemporary society. The poetry complicates the idea of utopia by suggesting that the correlation of a single ideology to place jars with more complex reality. This is best seen in Jamieson's poems in which attachment to the language and culture of rural Shetland sits comfortably and unsentimentally alongside global culture and concerns. What might be best seen as utopian in Scottish island poetry, then, is the potential of the localities it describes to inform wider debates about the ways we regard, value and make use of land-scapes and their ecosystems.

References

Appadurai, A., ed. 2001. *Globalization*. Durham, NC: Duke University Press.
Beer, G. 2003. "Island Bounds." In *Islands in History and Representation*, edited by R. Edmond and V. Smith, 32–42. London: Routledge.

Clark, T. 2012. "Scale: Derangements of Scale." In *Telemorphosis: Theory in the Era of Climate Change*. Vol. 1. edited by T. Cohen, 148–166. Ann Arbor, MI: Open Humanities Press.

Dunn, D. 2003. *New Selected Poems 1964–2000*. London: Faber and Faber.

Gifford, T. 1999. *Pastoral*. London: Routledge.

Hadfield, J. 2005. *Almanacs*. Highgreen: Bloodaxe.

Hadfield, J. 2008. *Nigh-No-Place*. Highgreen: Bloodaxe.

Heise, U. K. 2008. *Sense of Place and Sense of Planet: The Environmental Imagination of the Global*. Oxford: Oxford University Press.

Jameson, F. 1991. *Postmodernism, or, the Cultural Logic of Late Capitalism*. London: Verso.

Jamieson, R. A. 2007. *Nort Atlantik Drift*. Edinburgh: Luath Press.

Johnson, S., and J. Boswell. 1984. *A Journey to the Western Islands of Scotland and The Journal of a Tour to the Hebrides*. Edited by P. Levi. London: Penguin.

Longley, E. 2010. "Irish and Scottish 'Island Poems'." In *Northern Lights, Northern Words. Selected Papers from the FRLSU Conference, Kirkwall 2009*, edited by R. McColl Millar, 143–161. Aberdeen: Forum for Research on the Languages of Scotland and Ireland.

MacDiarmid, H. 1939. *The Islands of Scotland : Hebrides, Orkneys, and Shetlands*. London: B.T. Batsford.

MacDiarmid, H. 1993. *Complete Poems Volume I*. Manchester: Carcanet.

MacNeil, K., ed. 2011. *These Islands, We Sing: An Anthology of Scottish Islands Poetry*. Edinburgh: Polygon.

Moylan, T. 1986. *Demand the Impossible: Science Fiction and the Utopian Imagination*. New York: Methuen.

Paterson, D. 2003. *Landing Light*. London: Faber and Faber.

Quammen, D. 1996. *The Song of the Dodo: Island Biogeography in an Age of Extinctions*. London: Hutchinson.

Reid, A. 2008. *Inside Out: Selected Poetry and Translations*. Edinburgh: Polygon.

Robertson, R. 2010. *The Wrecking Light*. London: Picador.

Smith, I. C., ed. 1986. "Real People in a Real Place." In *Towards the Human: Selected Essays*, Edinburgh: MacDonald.

Smith, I. C. 2011. *Collected Poems*. Manchester: Carcanet.

Stewart, S. 1993. *On Longing: Narratives of the Miniature, the Gigantic, the Souvenir, the Collection*. Durham, NC: Duke University Press.

Thomson, J. 1981. *The Seasons*. Edited by J. Sambrook. Oxford: Clarendon Press.

Whittaker, R. J. 1998. *Island Biogeography: Ecology, Evolution, and Conservation*. Oxford: Oxford University Press.

Williams, R. 1973. *The Country and the City*. London: Chatto & Windus.

(Mis)Reading Hardy: 'Wessex' as green imaginary

Adrian Tait

Independent Scholar

The aim of this paper is to explore the parallels between Thomas Hardy's depiction of 'Wessex', and Kirkpatrick Sale's utopian vision of the bioregion. Although a fragmentary composite of the ideal and the actual, Hardy's Wessex can be (re)interpreted as a form of 'green imaginary', an idea that finds its fullest expression in *The Woodlanders*. Nevertheless, there are contrasts as well as similarities between Wessex and bioregion. In particular, Hardy's 'partly real, partly dream country' challenges both the viability and desirability of the bioregion as exclusionary entity, concerns that have grown steadily more relevant in the light of both globalised capitalism and a global environment crisis. However, Hardy's sense of place is itself more fluid and wide ranging than the (perhaps inevitable) association of Wessex with the local and parochial. Reread in terms of a process of imaginative reinvestment in the local *and* the global – a process that is most obvious in *The Dynasts* – Hardy's Wessex suggests a different and more adaptable kind of 'ecological thought'. It also raises serious questions about the extent to which human 'nature' is itself compatible with environmental ideals. Ultimately, therefore, the most ecotopian aspect of Hardy's Wessex may be its conventional, realist emphasis on human scale and human agency, and the promissory possibility that the individual can in the end make a difference.

From the moment that Thomas Hardy first used 'Wessex' to describe the world of his novels, it was both praised and derided as an ideal of what Shires calls the 'harmonious integration between man and nature' (Hardy 1874, xiii). It also made Hardy famous. What Wordsworth had done for the Lakes and the Brontës for the Yorkshire moors, remarked Lionel Johnson, Hardy had done for 'the land of his inventions' (1923, 83). But why did this 'partly real, partly dream country' exert such an influence (Preface, 1912, to *Far from the Madding Crowd*, in Orel 1966, 9)? For at least some of Hardy's newly urbanised contemporaries, Wessex represented the world they had just left behind, and to which they looked back with nostalgia and affection. Yet a century and more later, Wessex continues to beguile. Might it be that, from the outset, readers responded to 'Wessex' both as a compensatory pleasure *and* an anticipatory hope? As Ernst Bloch argued, 'the utopian impulse, the impulse to a better world, is ubiquitous in human culture; but its expression is necessarily historically variable, and often oblique and fragmentary' (*The Principle of Hope*, quoted in Levitas 2007, 53). Whilst it is self-evidently 'imperfect, incomplete, and non-totalizing' (Hardy 2012, 39), therefore, perhaps there is and always was an (eco) utopian dimension to Hardy's portrayal of 'Wessex'.

   This is not – or not simply – to exploit the definitional laxity of the utopian impulse, or the breadth, depth, and diversity of Hardy's body of work, which itself allows a multitude of very different readings. There are striking and precise parallels between Hardy's construct and late twentieth-century attempts to define a sustainable future. More specifically, Wessex can be read as a complex literary expression of Kirkpatrick Sale's bioregional vision.

The obvious difficulty lies in disentangling 'Wessex' from its familiar association with an idealised pastoral, a pastoral that has often been dismissed – and occasionally defended – as escapist. 'Escapist!' exclaimed Powys in his discussion of Hardy's work; 'that is the whole point' (1938, 621). Perhaps: or perhaps critics have misread as escapist what is in fact utopian, and in so doing, rewritten as reactionary what is in fact radical. This is not to ignore the backwards glance that Hardy repeatedly encodes in his depiction of an apparently settled, stable society, where the disruptive (Bathsheba, for example) are managed and outcomes (her eventual marriage to Gabriel Oak) reinforce the status quo. Nor is this to overlook the extent to which Hardy's is (in his own words) a 'modern Wessex of railways, the penny post, mowing and reaping machines, union workhouses, Lucifer matches, labourers who read and write, and National school children' (Preface, 1912, to *Far from the Madding Crowd*, in Orel, 9). *This* Wessex, Hardy insists, cannot escape its own historicity. But a disruptive surplus nevertheless remains.

How best to approach this process of rereading? Hardy's emphasis on the actual and historical allows Wessex to be seen in terms of 'both an *archaeological* or analytical mode, and an *architectural* or constructive mode' (Levitas 2007, 47). Rather than set an astringent critique of contemporary society against its utopian alternate, as William Morris did in *News From Nowhere*, Hardy offers a complex and continuous dialogue between the two. Wessex does not represent any single, settled, or stable state – no 'ecotopian' ideal – but rather, an evolving argument between what is possible and what is practical or plausible. Reflexive and provisional, it embodies Moylan's concept of a *critical* utopia, which deconstructs even as it reconstructs the tradition within which it is located (1986, 10); and if there is a strong parallel with the 'bioregional' project, we can therefore expect Hardy's own fragmentary projection of a sustainable utopia to challenge and undercut it.

But can the bioregional project itself be described as utopian? Writing at a time when the very word had become 'a chastisement' (1984, 245), Kirkpatrick Sale was himself insistent that the bioregional vision was less a fictional projection of a future society than a practical programme for the redirection of today's. Nevertheless, it is as 'an image of the future' ('positive and liberating') (245) that Sale's influence continues to be felt, most recently in Molly Scott Cato's discussion of a bioregional economy (2012, 6).

Here, it is useful to introduce a critical distinction between more conventional 'utopias of abundance', or 'technological utopias', and what de Geus calls 'utopias of sufficiency', or 'ecological utopias' (1999, 20–21), such as William Morris. Whilst contemporary society continues to pursue the 'utopia of abundance', in which happiness is equated with materialism, de Gues argues that an ecologically responsible society can flow only from the 'simplicity, self-restraint and moderation' (21) implied by its alternative, the utopia of sufficiency. Sufficiency is in turn embodied in the concept of 'sustainable development' (22) set out in Gro Harlem Brundtland's report on *Our Common Future* (1987). This is, however, only one of several, late twentieth century attempts to define a sustainable world. Perhaps the best known is the concept of the bioregion, which originated on the West Coast of America in the 1970s, and was later and provocatively reworked by 'the leading ideologue of bioregionalism', Kirkpatrick Sale (Cato 2012, 38).

The 'bioregion' is an area within which ecofootprint and geographical boundary are superimposed and mutually self-supporting, but also, as in Wessex, where personal and communal identity are intimately related to each other, and to physical setting. 'To become dwellers in the land', wrote Sale, 'to come to know the earth, fully and honestly, the crucial and perhaps only and all-encompassing task is to understand the place, the immediate, specific place, where we live' (224). According to Sale, the bioregion offers the opportunity to create stable, self-contained, and self-regarding communities, within which individuals are 'deeply bound together with other people and with the surrounding non-human forms of life in a complex interconnected web of being' (245–246). This, Sale argued, is a culture 'understandable because of its imminence in the simple realities of the surroundings' (245).

Those familiar with Hardy's work will immediately recognise – but also question – the parallel with Wessex. Sale's definition of the bioregion assumes that its 'rough boundaries are determined by natural rather than human dictates, distinguishable from other areas by attributes of flora, fauna, water, climate soils and landforms' (226). By contrast, Hardy's Wessex does not follow any single or obvious natural boundary. On the contrary: if we take the geological areas that Cato takes as a starting point for her discussion of 'provisioning bioregions' within the British Isles (31), it is immediately apparent that the Wessex of the novels (first illustrated in the edition of 1895–96) straddles several of them. There are vales, plains, hills, heaths, and forest, a natural diversity complemented and complicated by the human communities that helped shape them.

This is not, however, to invalidate the comparison with the bioregional construct. Whilst Sale wrote with the vast (and relatively empty) landmass of the United States in his mind's eye (227–279), he too described a hierarchy of 'bioregional gradations' (227), regions within regions 'narrowing in scale', and 'moving closer and closer to the specifics of the soil and those who live upon it' (229). Moreover, and notwithstanding his insistence on the importance of 'natural rather than human dictates' in shaping these gradations, he also accepted the defining role of 'the dwellers in the land, who will always know them best' (228). As Berg and Dasmann pointed out in one of the earliest statements of its principles, the 'final boundaries of a bioregion are best described by the people who have lived within it, through human recognition of the realities of living-in-place' (1977, 36). This is also to accept that the cultural and the natural inevitably interpenetrate. Seen in these terms, it becomes more apparent why Wessex can and should be read from a bioregional perspective, as the continuation (or culmination) of an ancient, Anglo-Saxon accommodation between the biotic and abiotic, a geo-political construct whose complexity is nevertheless a function of its natural diversity. It is the (bioregional) ideal read back into history, with provocative consequences.

This correspondence is at its most obvious in Hardy's *The Woodlanders*. 'Discrete and identifiable', the wooded world of the novel corresponds exactly with Sale's 'vitaregion' (227), the smallest of his gradations. Like the broader bioregional entity within which it nestles, this known and knowable community offers 'the satisfaction of being rooted in history, in lore, in place' (Sale 245); here, too, natural features have shaped a distinctive and responsive way of life (226). Nevertheless, the novel remains a complex hybrid, blending the ideal and the actual. This in turn generates the kind of critical dialogue to which I referred at the outset: the world of *The Woodlanders* both complements and challenges the bioregional vision and the utopian ideal it embodies.

The Woodlanders develops an idea suggested by the opening sentence of *Under the Greenwood Tree* ('to dwellers in a wood, almost every species of tree has its voice as well as its feature' [Hardy 1872, 11]), and rehearsed in *The Return of the Native*: the paramount importance of an 'intelligent intercourse with Nature' (Hardy 1887, 330). Those who live in Little Hintock have 'an almost exhaustive biographical and historical acquaintance with every object, animate and inanimate, within the observer's horizon' (1887, 125). None are more attuned to that environment than Giles and Marty:

> They had planted together, and together they had felled; together they had, with the run of the years, mentally collected those remoter signs and symbols which seen in few are of runic obscurity, but all together made an alphabet. From the light lashing of the twigs upon their faces when brushing them in the dark, they could pronounce upon the species of tree whence they stretched; from the quality of the wind's murmur through a bough, they could in like manner name its sort afar off. (331)

But whilst Giles and Marty may be the unaffected son and daughter of 'that wondrous world of sap and leaves called the Hintock woods' (330), whose value they recognise and whose meanings they understand without need of interpretation, the town educated have all too quickly forgotten what they once knew: as he returns with her from the railway station, Giles is surprised to find that Grace is no longer able to distinguish between different types of tree. Her mind is elsewhere, thinking of 'a broad lawn in the fashionable suburb of a fast city' (42). In 'all the pride of life', she has 'fallen from the good old Hintock ways' (42, 44). The town-bred, on the other hand, cannot forget an 'old association' (125) that they never had. As the narrator points out, '[t]he spot may have beauty, grandeur, salubrity, convenience; but if it lacks memories it will ultimately pall upon him who settles there without opportunity of intercourse with his kind (125)'. 'I was made for higher things', declares Fitzpiers, between yawns (49). Felice is similarly characterised by a 'mien of listlessness' (59). Unable to appreciate the significance of what lies within her own horizon, she travels the world gathering surface impressions that never cohere: 'I think sometimes I was born to live and do nothing, nothing, nothing but float about', she declares (59). There are dangers in the narrow parochialism of a closed community, 'where reasoning proceeds on narrow premises', but these citizens of the world are no happier for their grand tours (9).

In part, therefore, *The Woodlanders* appears to embody the schematic contrast between the values of town and country or, as Ryle remarks, between a 'developed metropolitan consciousness and mutely autochthonous "organic" being' (2002, 19). It is also the contrast between the values of a restless, modern existence, and those of a life lived in environmental balance, a balance about which the absentee landowner, Felice, knows little and cares less: 'I might fell, top, or lop, on my own judgment, any stick o' timber whatever in her wood', declares Melbury; 'I wish she took more interest in the place, and stayed here all year round' (46).

The parallels with Sale's bioregional vision are immediately apparent. These are, however, the very aspects of the bioregional vision that have tended to attract criticism. For example, the bioregional vision is often said to suffer from an anti-urban bias, a bias that is, at best, self-defeating, in a world increasingly made up of city dwellers. Apologists point out that there is nothing inherently anti-urban in a movement that has also 'fostered "green city" efforts' (Lynch, Glotfelty, and Armbruster 2012, 7), and Hardy's Wessex, whilst still largely dominated by field and hedgerow, nevertheless makes room for (small) towns that sit comfortably with a developed bioregional vision. This is a view shaped by a sense of symbiosis, not of dominance, where, as in *The Mayor of Casterbridge*:

> Bees and butterflies in the cornfields at the top of town, who desired to get to the meads at the bottom, took no circuitous course, but flew straight down High Street without any apparent consciousness that they were traversing strange latitudes. (Hardy 1886, 88)

Furthermore, *The Woodlanders* is itself more subtle than the schematic conflict between town and country, or a simplistic account of 'the impact of an urban alien on the "timeless pattern" of English rural life' (Williams 1984, 102). Country life has generated its own pressures, its own catalysts for change: for Grace Melbury, as for Tess, Clym, and Jude, a supposed rural idyll is undone from within, not from without, by a growing awareness of other, 'incompatible ways of being' (Ryle, 22). This is a point that Hardy made forcibly in his essay on 'The Dorsetshire Labourer', in which he suggested that, inevitably, urbanised existence appeared more appealing to those who lived and worked on the land; and if that was not to the taste of middle-class outsiders, well, 'it is only the old story that progress and picturesqueness do not harmonise': 'it is too much for them to remain stagnant and old-fashioned for the pleasure of romantic spectators' (Orel, 181). As Tess protests (Hardy 1891, 232), 'I am only a peasant by position, not by nature!'

This sense of change at work from within nevertheless contrasts, not only with the conventional view of Wessex as a stable, settled society, but with Sale's assumption that 'nature is inherently stable' (229). In fact, any 'balance' or 'steady state' is temporary, illusory, or artificial, as Botkin points out in *Discordant Harmonies*. '[P]sychologically uncomfortable' as it may be to abandon a long-standing belief 'in the constancy of undisturbed nature', Botkin notes, we are in fact 'a part of a living and changing system' (1990, 188, 189). In Hardy's post-Darwinian world – the world of Little Hintock – there is a strong sense of a restless and dynamic, but also deeply disconcerting 'nature', in which stoats are glimpsed 'sucking the blood of the rabbits', a sunless winter's day dawns like 'the bleared white visage' of 'a dead-born child', and leaves are 'dwarfed and sickly for want of sunlight' (1887, 23, 25). It is very far from 'ordered, purposive, benign' (Richards 2001, 154). The very trees are 'wrestling for existence' (Hardy 1887, 311).

That struggle finds its inevitable analogy in the (hard) lives lived by Hardy's characters. Following Sale (230), it can be argued that, in Wessex, resource use is minimised, conservation emphasised, and waste avoided. Self-sufficiency – that most elegant of principles – is indeed its basis (Sale, 230). But by any standard, life in Little Hintock is difficult, and Hardy does nothing to conceal the fact.

The novel's opening description leads immediately to Marty (Chapter 2), sat making spars in the cottage where her father lies sick. To make a living, she must do the work he cannot, and works until three in the morning (Chapter 3) to do so, her hands 'red and blistering' (10). Furthermore, she must keep it a secret that the work is hers; this is a trade for which she is not and cannot as a woman be qualified, despite her natural aptitude (22). Not that it is enough to maintain her. To make ends meet, she is forced to sell her hair for a sovereign it would otherwise take a week and a half to earn 'at that rough man's work' (12). Although 'struggling bravely', her efforts may in any case come to naught: with the death of her father, she will lose her home (46). In the world of Little Hintock, therefore, self-sufficiency is less a lifestyle choice than a function of a fundamental and grinding poverty.

This in turn highlights another critical contrast between Hardy's Wessex, and Sale's bioregional vision. Sale assumes that a bioregion can exist in splendid, self-contained isolation. Hardy's Wessex suggests otherwise. Sequestered as Little Hintock may be, it has already been assimilated into a wider 'web of human and technical relations' that both creates and exploits inequality, as Marty's poverty demonstrates (Watson 1996, 144). For whilst Marty's relationship to her environment may be intimate and immediate, others have already appropriated her labour. She does not own the wood from which she makes spars, since there are no commons left from which to gather it; she does not make those spars to fix her own roof, since she has no roof of her own, and will have no home the moment her father dies and the life-hold lease expires. She exists, or rather, subsists, only because of the monetary value accrued by her labour. Her relationship to the environment is everything a bioregionalist might wish for, but it does not survive first contact with an already globalised capitalist economy that has bound its citizens in wage slavery and further discriminated against them by gender. Marty is a marginal presence in the world of the woodlanders; indeed, she is a marginal presence in the novel itself. Ultimately, this is true also of Giles. They may be authentic representatives of a so-called 'organic' community, but it has already been overwhelmed, as have they. Capitalism is corrosive and, as Kerridge remarks, Giles' and Marty's 'close association with trees signifies deep-rootedness but also vulnerability' (2000, 137): theirs is a world as fragile as any ecosystem, delicately poised, and susceptible to upset.

It is, in fact, impossible to conceive of a community that is not in some way embedded in what Heise calls 'global networks of information and exchange' (2008, 54). This leads, in turn, to the 'deterritorialization of local knowledge' (Heise, 55), and even in Hardy's lifetime, the loss of what he called 'a vast mass of unwritten folk-lore, local chronicle, local topography, and nomenclature': 'there being no continuity of environment', he added, 'there is no continuity of information' (Millgate 1984, 336).

There is a related question: whether or not a genuinely self-contained and self-sufficient bioregion is viable, is it actually desirable? Manifestly, environmental crisis also involves problems that can only be addressed through systematic global intervention. At the very least, these must include 'policies of redistribution and cost-sharing' and a concerted drive towards the consistent ratification of international environmental law (George and Wilding 2002, 198). How can a sense of place of the kind that Hardy describes and Sale imagines, delimited and specific, ever be adequate to meet problems on a worldwide scale? As Heise contends, 'the challenge for environmentalist thinking, then, is to shift the core of its cultural imagination from a sense of place to a less territorial and more systematic sense of planet' (56). Her answer is *eco-cosmopolitanism*, the attempt to 'envision individuals and groups as part of planetary "imagined communities" of both human and nonhuman kinds' (61).

Bioregionalists have, in turn, countered that this is a 'matter of emphasis', sensing that 'the division of the local and global is a false dichotomy that limits the possibilities for imagining environmentally responsible global citizenship' (Lynch 2012, 9, 10). And perhaps paradoxically, Hardy does not offer the 'either/ or' of place against planet. The utopian promise of Wessex in part lies in its refusal to see any opposition between the two, or rather, to suggest that, as Murphy remarks, the local can 'provide the experiential basis for the appreciation of the global' (2009, 41).

What, then, does Hardy offer?

For its proponents, the central challenge of the bioregional vision – and perhaps its 'most distinctive key term' – is re-inhabitation (Lynch 2012, 6). In one of its earliest statements, however, Berg and Dasmann defined a bioregion as both 'a geographical terrain and a terrain of consciousness' (1977, 36). This is a terrain that, by definition, can be (mentally) mapped on to any environment (whether urban, rural, or edgeland, or whether local or global). It is, in effect, a process of imaginative reinvestment. If humankind cannot go 'back to the land' – a project doomed even in Hardy's time, as Gould (1988) notes – it can re-engage with its environment psychologically, and, in so doing, reach beyond the constraints of place to a much more comprehensive sense of planet. And if, as Buell remarks, environmental crisis is a 'crisis of the imagination', then this process of mental mapping is not simply desirable, but essential (1995, 2).

Hardy offers his own version of this process. Early in *Far from the Madding Crowd*, the narrator describes the night sky above Norcombe:

> To persons standing alone on a hill during a clear midnight such as these, the roll of the world eastward is almost a palpable movement. The sensation may be caused by the panoramic glide of the stars past earthly objects, which is perceptible in a few minutes of stillness; or by the better outlook upon space that a hill affords, or by the wind; or by the solitude; but whatever be its origin the impression of riding along is vivid and abiding. (1874, 15)

In moments of sensuous specificity like this one, Hardy reconnects reader and environment (and as Naess remarks, 'ethics *follow from* how we experience the world' [1989, 20]). Indeed, it might profitably be re-read as a concrete expression of Heidegger's Dasein. Translated not as 'being' but as 'being there', Dasein implies just the kind of immediacy – but also that expanded sense of 'world' – that the narrator describes and the reader shares.

Heidegger's influence can be felt on Naess, as Heise points out (35), and, 'once – or twice-removed', on Sale, for whom 'to dwell' means 'to be fully engaged to the sensory richness of our immediate environment' (Lynch 2012, 5). Bioregions can, therefore, be regarded 'as more phenomenonologically real than politically constructed places' (Lynch 2012, 5). But Hardy's is not, transparently, an unreflective engagement with the environment, but a complex and carefully calculated literary performance.

It was necessarily so. Whether or not 'ties to the natural world' inevitably flow from a close and intimate relationship with a particular place, as environmentalists may sometimes assume (see Heise, 61), Hardy's 'Wessex' was not aimed at readers who still had the good fortune – or otherwise – of that kind of relationship. *His* audience was always predominantly urban: at the time that *Far from the Madding Crowd* made his name, two-thirds of the population of England and Wales already lived in towns and cities (Read 1994, 46). As Buell notes, Hardy was himself a 'semi-deracinated returnee' (2005, 96), in Hardy's own words, 'vibrating at a swing between the gaieties of a London season and the quaintnesses of a primitive rustic life' (Millgate 1984, 257). If there is a strong vein of nostalgia in Hardy's depiction of Wessex, it is simply because the actual world of work from which it derives was 'almost extinguished' by 1912, as Hardy remarked in a later Postscript to *The Woodlanders* (1887, 369). What he offered his readers was, instead, the opportunity to immerse themselves in its imagined equivalent; a 'community' built on a human scale, and structured around shared feelings, shared knowledge, and shared values. In short, he offered a literary (if not a literal) act of re-inhabitation.

As Heise notes, the challenge of 'environmentalist advocacies of place' is their assumption that 'existential encounters' can be 'recuperated intact from the distortions of modernity' (54). Not the least of those distortions is the fact that, today, 'whatever knowledge inhabitants acquire about a particular place is for the most part inessential for their survival' (Heise, 55). This cannot be said of Marty and Giles and their own experience of place, but it is exactly the point that Hardy makes in describing *The Return of the Native*. As the narrator dryly remarks of Clym Yeobright, the returning native of the title, a 'man who advocates aesthetic effort and deprecates social effort is only likely to be understood by a class to which social effort has become a stale matter' (1878, 175).

And a close reading of his oeuvre suggests that Hardy's intention was never to rescue a sense of place intact from the 'distortions of modernity' that Heise describes, but rather, to exploit those distortions as a liberating antidote to the narrowness of local perceptions, and open up a horizon that encompassed the global as well as the local. This is already apparent in Hardy's carefully staged account of the night sky above Norcombe Hill, which owes its power not to any pre-reflective moment of communion with the universe, but to its mediation through a detailed account of setting, the memorable power of metaphor ('the roll of the world eastward'), a scientistic (if not scientific) description of the night sky ('a difference of colour in the stars' …), and to its strategic placement both early in the chapter and in the novel.

These carefully contrived and constructed passages recall Shklovsky's dictum, that 'art exists that one may recover the sensation of life; it exists to make one feel things, to make the stone *stony*' (1917, 12). And if, as Morton argues, 'ecological art' is as important for what it *does* as for what it *is* (2010, 11), this is critical to our understanding of the imaginative but also processual feat that Wessex represents, and to its later expression in Hardy's verse and verse-drama, (in)famous for its stylistic and formal difficulties. Here, it becomes all the more apparent that, whilst Sale's vision of the bioregion is explicitly opposed to cosmopolitanism (224), Hardy's critical utopia was always open to what Buell calls 'translocal forces' (2005, 88). Whether glancing outwards to the aftermath of colonial (mis)adventure in 'Drummer Hodge', inverting (inward-looking) localism in 'Geographical Knowledge', or questioning the inevitability – the immutability – of national boundaries in 'His Country', Hardy widens 'the circles of place' (Buell 2005, 93).

> I traced the whole terrestrial round,
> Homing the other side;
> Then said I, 'What is there to bound
> My denizenship? It seems I have found
> Its scope to be world-wide.
> ('His Country', ll.16–20, in Gibson 2001, 539)

This spacious perspective finds its most remarkable expression in *The Dynasts* (1903–1908), Hardy's epic account of the Napoleonic War. Although Wessex remains central to the drama, *The Dynasts* looks outwards. Scene by scene, the action switches from place to place, country to country, as Hardy extends the hawk-like perspective of earlier poems like ' "In vision I roamed" '. Seen from 'mid-air', the Danube 'shows itself as a crinkled satin riband' (Hynes 1995a, vol. IV, 62). By night, London and Paris 'seem each as a little swarm of lights surrounded by a halo' (Hynes vol. IV, 197). Europe is 'disclosed as a prone and emaciated figure, the Alps shaping like a backbone, and the branching mountain-chains like ribs' (Hynes vol. IV, 20). But the human narrative is only one part of the tale that Hardy has to tell. Whilst 'monumental and universal', notes Hynes, Hardy's epic verse drama is also 'attentive to existence on the smallest possible scale' (vol. IV, xxiii). The subject of *The Dynasts*, wrote White, is 'never simply Europe, or mankind': it is 'all sentient life' (1974, 129).

Form is again critical to Hardy's imaginative achievement. Here, he creates a 'verse-drama' that blends prose and poetry; as a 'non-realist' text (Widdowson 1996, 77), its style, shape and structure place it amongst the most radical works in the English language. This in turn enables Hardy to introduce and sustain the audacious idea that binds this three-volume epic together, an idea that fully embraces 'the radical openness' of Morton's 'ecological thought' (11). As the spirits look down on the drama's 'Fore Scene', 'a new and penetrating light' reveals 'as one organism the anatomy of life and movement in all humanity and vitalized matter' (Hynes vol. IV, 21). This is 'the Immanent Will' whose 'strange waves', 'twining and serpentining round and through' 'complicate with some, and balance all' (Hynes vol. IV, 21). Lacking 'right or reason' (Hynes vol. IV, 56), it nevertheless weaves every life form into a single web, at once dissolving 'local' and 'global'. More: by insisting on the interrelatedness of all life, the Immanent Will establishes the basis for an environmental ethic that recognises and accepts that intrinsic value is freely distributed throughout the environment.

Hardy's vision of 'radical coexistence' anticipates Morton's 'mesh' (10, 8), but it also has its analogy in James Lovelock's 'gaia', the hypothesis that 'the earth's living matter, air, oceans, and land surface form a complex system which can be seen as a single organism and which has the capacity to keep our planet a fit place for life' (quoted in Dobson 1991, 265). Gaia itself has pride of place in Sale's bioregional vision (223), both in the sense that Lovelock has now given it, and as an ancient, animistic conception (to quote Sale) of the earth as single organism, 'self-contained and almost purposeful' (219). For Sale, the success of his utopian project depends on recovering this sense of the earth as 'a living being, sentient and organic' (221), and recognising that humankind is a part of it.

And *The Dynasts* embodies just this (startling) possibility. In its closing pages, Hardy suggests that a collective consciousness might one day animate the 'web Enorm', 'till It fashions all things fair!' (Hynes 1995b, vol. V, 252, 255). Yet the Immanent Will, as Hardy elsewhere stresses, is not set apart from the mass of humankind: it reflects 'a sense that humanity and other animal life' form 'the conscious extremity of a pervading urgence, or will' (Millgate 2001, 200):

> Will that wills above the will of each,
> Yet but the will of all conjunctively.
> (Hynes vol. V, 31)

Perhaps, then, its growing consciousness may prompt a shared realisation that individuals are inevitably enmeshed in the collective; perhaps 'consciousness, and responsibility, will come to the process of – and the players in – history' (Richardson 2004, 173).

Morton demands that the ecological thought think big; no idea could be bigger. And yet, even as Hardy's remarkable concept of the Immanent Will appears to anticipate and even transcend the most utopian of Sale's arguments, his view of life complicates the comparison. Notwithstanding the optimistic note on which it concludes, it is one of the more frightening implications of *The Dynasts* that the destiny of humankind is fixed, and fixed by pre-determined patterns of behaviour. In Hardy's epic-drama, humans act in a mass, 'writhing, crawling, heaving, and vibrating in their various cities and nationalities', thrust 'hither and thither' by the Immanent Will (Hynes vol. IV, 20, 160).

This denial of human agency – this radical decentring of the human – may in part explain *The Dynasts'* failure to capture the imagination of readers steeped in a liberal literary tradition built on the idea of free will, and of individual importance. Humans would rather dream:

> Their motions free, their orderings supreme;
> Each life apart from each, with power to mete
> Its own day's measures; balanced, self-complete;
> (Hynes vol. IV, 21)

But that dream is, perhaps, a final, rather contrary reason why Hardy's work speaks to the utopian tradition, and no less to the ecotopian project. Utopian visions have often been criticised for the way in which they impose uniformity on their subjects, as Popper argued in his critique of Plato's 'totalitarian moral theory' (1966, 119). As Berneri notes, 'utopian men are uniform creatures with identical wants and reactions and deprived of emotions and passions, for these would be the expressions of individuality' (1987, 5). By contrast, Arendt (1958) argues that it is individuality that makes change possible:

> Action would be an unnecessary luxury, a capricious interference with general laws of behaviour, if men were endlessly reproducible repetitions of the same model, whose nature or essence was the same for all and as predictable as the nature or essence of any other thing. Plurality is the condition of human action because we are all the same, that is, human, in such a way that nobody is ever the same as anyone else who ever lived, lives, or will live. (8)

And it may be that, after all, Hardy is at his most idealistic – his most utopian – not in those innovatory works that, like *The Dynasts*, de-centre our belief in the primacy of the individual, but in those realist constructs – the novels – which do the opposite, and (ultimately) assert an individual's importance. Confronted by the overwhelming scale of environmental crisis – indeed, the very word 'environment' may seem unimaginably vast and therefore personally meaningless – there is an entirely natural tendency to assume that the actions of any one individual can never make a difference. No one individual (and therefore no-one) can affect the outcome. But, and despite the complex forces that shape and affect their actions, the stories that Hardy tells are about individuals. Although invariably at odds with the forces that (mis)shape their destinies, they fashion their own futures. And even in the face of the great apocalypse of his time (a war poignantly called 'the war to end all wars'), Hardy maintained that humankind might still possess a 'modicum of free will' with which to decide its own fate, and shape for itself a better and more peaceful future ('Apology' to the *Late Lyrics and Earlier*, in Gibson 2001, 558). As the Chorus of the Pities sings (in 'aerial music') to the Shade of the Earth:

> We would establish those of kindlier build,
> In fair Compassions skilled,
> Men of deep art in life-development;
> Watchers and warders of thy varied lands,
> Men surfeited of laying heavy hands
> Upon the innocent,
> The mild, the fragile, the obscure content
> Among the myriads of thy family.
> (Hynes vol. IV, 16–17)

In conclusion, how does Hardy's 'Wessex' compare (and contrast) with the bioregional ideal, and Kirkpatrick's Sale's version of it? There are (sometimes startling) coincidences and correspondences, but they are throughout complicated by Hardy's awareness of wider historical forces and a profoundly cautious – some would say pessimistic – view of human nature. Wessex is, as I have suggested, a critical utopia: it refers back to the moment of its own dissolution as well as forward to the possibility, perhaps even the necessity, of its recreation. To that extent, it stands in opposition to (and often undercuts) Sale's own vision of a sustainable future, with its confident depiction of human societies shaped by the land in which they have been (re-)embedded. In the great novels of what Hardy himself called 'character and environment', by contrast, people and place co-exist in a temporary state of tension, in which relationships must be constantly re-negotiated. And if Wessex is a utopia by default, rather than by design, this may be its strength. Critics from Burke to Popper have protested that utopian projects are by their nature irrational and, in their desire to force a radical break with history and tradition, dangerous. It matters not, wrote Popper, whether the utopian impulse seeks its 'heavenly city in the past or in the future', or whether it preaches ' "back to nature" or "forward to a world of love and beauty" ' (1966, 168). There can be no definitive ideal, only 'trial and error', an evolving accommodation, a piecemeal rather than 'complete reconstruction of our social system': it is a matter, wrote Popper, of inspiration 'checked by experience' (167). But an inspiration it remains. As Dolin remarks, Hardy's 'life and work show us how, through the power of the imagination, we can connect meaningfully and without nostalgia with *where we are* – with real and virtual landscapes, and local, national and global histories and memories' (2008, 9). In a world now dominated by the city and in a human society increasingly distanced from everyday encounters with the non-human, that is, perhaps, the most meaningful but also the most idealistic of bioregional constructions: it is one that, even if ruefully, accepts and negotiates the continuing transition from a closed and so-called 'organic' community to an open and abstract society.

References

Arendt, H. 1958. *The Human Condition*. 2nd ed. Chicago, IL: University of Chicago Press.
Berg, P., and R. Dasmann. [1977] 1990. "Reinhabiting California." In *Home! A Bioregional Reader*, edited by V. Andruss, C. Plant, J. Plant, and E. Wright, 35–38. Philadelphia, PA: New Society.
Berneri, M. L. 1987. *Journey Through Utopia*. London: Freedom Press.
Botkin, D. B. 1990. *Discordant Harmonies: A New Ecology for the Twenty First Century*. New York: Oxford University Press.
Buell, L. 1995. *The Environmental Imagination: Thoreau, Nature Writing, and the Formation of American Culture*. London: Belknap Press.
Buell, L. 2005. *The Future of Environmental Criticism: Environmental Crisis and Literary Imagination*. Oxford: Blackwell.
Cato, M. S. 2012. *The Bioregional Economy: Land, Liberty and the Pursuit of Happiness*. Oxford: Routledge.
de Geus, M. 1999. *Ecological Utopias: Envisioning the Sustainable Society*. Utrecht: International Books.
Dobson, A., ed. 1991. *The Green Reader*. London: André Deutsch.
Dolin, T. 2008. *Thomas Hardy*. London: Haus Publishing.
George, V., and P. Wilding. 2002. *Globalization and Human Welfare*. Basingstoke: Palgrave.
Gibson, J., ed. 2001. *Thomas Hardy: The Complete Poems*. Basingstoke: Palgrave.
Gould, P. C. 1988. *Early Green Politics: Back to Nature, Back to the Land, and Socialism in Britain 1880–1900*. Sussex: Harvester.
Hardy, K. 2012. "Unsettling Hope and Re-articulating Utopia." *Puerto del Sol* 47 (2): 36–43. Accessed March 17, 2013. http://www.puertodelsol.org/assets
Hardy, T. [1872] 1992. *Under the Greenwood Tree*. Edited by Simon Gatrell. Oxford: Oxford University Press.
Hardy, T. [1874] 2002. *Far from the Madding Crowd*. Edited by Suzanne B. Falck-Yi, with an introduction by Linda M. Shires. Oxford: Oxford University Press.

Hardy, T. [1878] 1998. *The Return of the Native*. Edited by Simon Gatrell. Oxford: Oxford University Press.

Hardy, T. [1886] 2004. *The Mayor of Casterbridge*. Edited by Dale Kramer. Oxford: Oxford University Press.

Hardy, T. [1887] 1998. *The Woodlanders*. Edited by Patricia Ingham. London: Penguin.

Hardy, T. [1891] 2003. *Tess of the D'Urbervilles*. Edited by Tim Dolin. London: Penguin.

Heise, U. 2008. *Sense of Place and Sense of Planet: The Environmental Imagination of the Global*. Oxford: Oxford University Press. doi:10.1093/acprof:oso/9780195335637.001.0001.

Hynes S., ed. 1995a. *The Complete Poetical Works of Thomas Hardy: Volume IV*. Oxford: Clarendon Press.

Hynes S., ed. 1995b. *The Complete Poetical Works of Thomas Hardy: Volume V*. Oxford: Clarendon Press.

Johnson, L. [1894] 1923. *The Art of Thomas Hardy*. London: John Lane.

Kerridge, R. [2000] 2001. "Ecological Hardy." In *Beyond Nature Writing: Expanding the Boundaries of Ecocriticism*, edited by K. Armbruster, and K. R. Wallace, 126–142. London: University Press of Virginia.

Levitas, R. [2007] 2009. "The Imaginary Reconstitution of Society: Utopia as Method." In *Utopia, Method, Vision: The Use Value of Social Dreaming*, edited by R. Bacconlini, and T. Moylan, 46–68. Bern: Peter Lang.

Lynch, T., C. Glotfelty, and K. Armbruster, eds. 2012. *The Bioregional Imagination*. London: University of Georgia.

Millgate, M., ed. 2001. *Thomas Hardy's Public Voice*. Oxford: Clarendon Press.

Millgate, M., ed. [1984] 1985. *The Life and Work of Thomas Hardy*. Basingstoke: Macmillan.

Morton, T. 2010. *The Ecological Thought*. London: Harvard University Press.

Moylan, T. 1986. *Demand the Impossible: Science Fiction and the Utopian Imagination*. London: Methuen.

Murphy, P. D. 2009. *Ecocritical Explorations in Literary and Cultural Studies; Fences, Boundaries, and Fields*. Lanham, MD: Lexington.

Naess, A. 1989. *Ecology, Community and Lifestyle*. Translated and revised by David Rothenberg. Cambridge: Cambridge University Press. doi:10.1017/CBO9780511525599.

Orel, H. 1966. *Thomas Hardy's Personal Writings*. Lawrence: University of Kansas Press.

Popper, K. R. [1966] 1969. *The Open Society and Its Enemies: Volume 1, The Spell of Plato*. 5th ed. London: Routledge & Kegan Paul.

Powys, J. C. 1938. *The Pleasures of Literature*. London: Cassell.

Read, D. 1994. *The Age of Urban Democracy: England 1868–1914*. London: Longman.

Richards, B. 2001. *English Poetry of the Victorian Period, 1830–90*. 2nd ed. London: Pearson.

Richardson, A. 2004. "Hardy and Science: A Chapter of Accidents." In *Palgrave Advances in Thomas Hardy Studies*, edited by P. Mallett, 156–180. Basingstoke: Palgrave Macmillan.

Ryle, M. 2002. "After 'Organic Community': Ecocriticism, Nature, and Human Nature." In *The Environmental Tradition in English Literature*, edited by J. Parham, 11–23. Aldershot: Ashgate.

Sale, K. [1984] 1986. "Mother of All: An Introduction to Bioregionalism." In *The Schumacher Lectures: Second Volume*, edited by S. Kumar, 219–250. London: Abacus.

Shklovsky, V. [1917] 1965. *"Art as Technique." Russian Formalist Criticism*. Translated by Lee T. Lemon and Marion J. Reis. Lincoln: University of Nebraska Press.

Watson, D. 1996. *Beyond Bookchin: Preface for a Future Social Ecology*. Detroit: Black & Red.

White, R. J. 1974. *Thomas Hardy and History*. London: Macmillan.

Widdowson, P. 1996. *Thomas Hardy*. Plymouth: Northcote House.

Williams, R. [1984] 1987. *The English Novel from Dickens to Lawrence*. London: Hogarth.

Utopian Zionist development in Theodor Herzl's *Altneuland*

Nicola Robinson

Department of English and Related Literature, University of York, York, UK

Theodor Herzl's *Altneuland* depicts the opening stages of the Zionist settler-colonial project in Mandate Palestine at the beginning of the twentieth century. Specifically, *Herzl* represents land use, agricultural labour and Zionist collectivism, in other words the material means by which settler-colonisation can be achieved. This essay examines utopia and other literary strategies including *Bildungsroman* to demonstrate the narrative trajectory's move away from an underdeveloped environment and society towards a prosperous one. This essay argues that in *Altneuland* Herzl projects a vision of settler-colonialism that is different to the contemporaneous reality, such that the text can be regarded as an alternative to the real-world 'utopian' project of Israel/Palestine.

Theodor Herzl's *Altneuland* (published in English as *Old New Land*) depicts the opening stages of the Zionist settler-colonial project in Mandate Palestine at the beginning of the twentieth century. The text was originally published in German in 1902, although set 20 years in the future. *Altneuland* represents land use, agricultural labour and Zionist collectivism, in other words the material means by which settler-colonisation can be achieved. Graham Huggan and Helen Tiffin assert that 'postcolonial ecocriticism [has the] capacity to set out symbolic guidelines for the material transformation of the world' (2010, 14). I offer a postcolonial[1] ecocritical reading of *Altneuland* that demonstrates the way Herzl's narrative trajectory moves away from an underdeveloped environment and society towards a prosperous one. Herzl's promotion of settler-colonialism is encapsulated in his use of utopia and *Bildungsroman*. This essay looks at two utopian projects, one literary and one geopolitical, and how the constraints and possibilities of each of these utopias conflict. In particular, Herzl's use of the generic characteristics of utopian fiction projects a vision of settler-colonialism that is different to the contemporaneous reality. This essay argues that these literary strategies enable Herzl to create a text that can be regarded as an alternative to the real-world 'utopian' project of Israel/Palestine.

The essay begins with a brief background to the historical origins of development discourse and practice in the decades preceding the establishment of the state of Israel in 1948 because it is essential for contextualising the novel's representation of its interdependence with settler-colonialism and nation-building. Development in this context refers to the discourses that were conceptualised by the founders and intellectuals of Labour Zionism in order to establish a self-sufficient Jewish community and ensure its economic survival. I also account for Herzl's political formation, specifically his development of Zionism which provides evidence of his intentionality in the novel. Prior to a close reading of *Altneuland*, I also examine how the novel corresponds to and departs from the generic characteristics of utopian fiction.

Labour Zionism and Zionist development

Labour Zionism 'gave precedence to the realisation of an ethnocratic settler project: the establishment in Palestine of a mono-religious, ethnic settler state... [and] collective (Jewish) control of the land' (Masalha 2007, 25). Settlement and agricultural colonisation[2] were to be put into practice by the members of the Yishuv[3] in order to begin the task of establishing a nation state in the perceived Jewish homeland of Eretz Yisrael. Eretz Yisrael is both a theological and an ecological concept. I term 'Zionist development' the Jewish people's return to and settlement in the land of their ancestors and the control and cultivation of this land through the labour of the Yishuv. Derek Penslar aptly observes that '[a]s a European nationalist movement, Zionism could not help but have a powerful pedagogic and developmental dynamic' (2001, 94). As such, for Jews, Zionism adopted an educational role by incorporating nationalism and colonialism. Settler nationalism was infused with a vision of agricultural colonisation as a return to the Biblical homeland of the Jewish people. Arthur Ruppin, in the Preface to his *The Agricultural Colonisation of the Zionist Organisation in Palestine*, is explicit about the interconnection of these two elements: 'without agricultural colonisation the construction of a Jewish National Home in Palestine is an impossibility, and that for this reason we dare not shirk the struggle with the difficulties, nor abandon it till we have successfully overcome them' (1976, vi). 'The *Palestine Office* was established in Jaffa in 1908 by Ruppin as the representative of the *World Zionist Organization* in Palestine. Over the decade, the PO was the central agency for Zionist settlement activities in Palestine. ... The *Palestine Land Development Company* was established in 1909 by Ruppin and Otto Warburg as an instrument for the acquisition and development of land in Palestine' (Bloom 2011, 1–2). Drawing on Anthony Vital's observation that ecocriticism must be 'rooted in local (regional, national) concerns for social life and its natural environment' (2008, 88), I argue for a reading of *Altneuland* which draws attention to how the Yishuv can return to and settle on the land as well as how they can cultivate or 'redeem' the land through their agricultural labour.

The historical basis of Israel's social and economic development relates to the several types of agrarian settlement that were established in Palestine from the late nineteenth century. The labour or settler-colonial dimension of Zionism was realised through the members of the Yishuv, who are often referred to as the 'practical Zionists' or 'pioneers', and their Aliyah or immigration to Palestine. Gudrun Kramer acknowledges the influence of the Bible: 'the Hebrew term '*aliya* was a moral term for immigration to Eretz Israel, defining it not as mere migration, but rather as an "ascent"...to the Jerusalem Temple, standing "up on the mountain" ' (2008, 104). Although agricultural production remained the model of economic growth and self-sufficiency for a collective community, the settlement format and conceptualisation of labour differed over the successive *aliyot*.

Theodor Herzl and Zionism

Theodor Herzl was born in Budapest in 1860. After graduating from the University of Vienna with a doctorate in law, he wrote a number of texts exploring possible solutions to the 'Jewish Question' of anti-Semitism throughout Europe. Herzl is widely regarded as the founding father of political Zionism which is concerned with creation of a sovereign Jewish state in Palestine. His Zionist vision and project of bringing together a Jewish national community with a project of state-building was presented in its entirety in a book entitled *Der Judenstaat* (1896, and translated and published into English as *A Jewish State*). The following year, Herzl convened the First Zionist Congress in Basel, Switzerland. At this conference, the World Zionist Organization was founded and Herzl was made its first president. *Altneuland* is Herzl's first fictional publication and was originally published in German in 1902, although set 20 years in the future. The text was translated into English in the same year as *Old New Land* in a series of instalments for the US journal *The Maccabean*, a monthly English-language Zionist magazine, published by the Federation of American Zionists in New York. The narrative depicts the ideological and practical aspects of how settlement and agricultural colonisation can be achieved in Palestine by the Jewish people.

In July 1898, Herzl gave an interview to the *Young Israel* journal in which he invoked the text of *Robinson Crusoe* and aligned it to the fulfilment of the Zionist project in his novel *Altneuland*. He observes that '[a]ll the means we need, we ourselves must create them, like Robinson Crusoe on the island – your readers will surely understand the hint. In the days to come, the story of Zionism's development will be like a wonderful novel' (cited in Eitan Bar-Yosef 2007, 87). Herzl refers to the rise of the novel and its role in creating national consciousness and a national community (Anderson 1991, 25) as well as the specific text that will serve Zionism as the paradigm of conquest, development and control of land and labour. *Altneuland* recounts the travels of a young Jewish lawyer, Friedrich Loewenberg, and his travelling companion Kingscourt who decide to remove themselves from society by embarking on a long sea voyage. Over the course of this journey, they make two visits to Palestine: once in 1902 and again in 1923. In this sense, Herzl's novel corresponds to the conventions of utopia, whereby the setting is usually toured by a visitor, whereas dystopias are depicted from the point of view of an inhabitant. Additionally, *Altneuland* adheres to two main generic characteristics of utopian fiction: weak characterisation and limited plot development. In what follows, I explore how the form of utopia is used by Herzl to depict an unrealised future. His narrative traces a trajectory away from underdevelopment towards prosperity in order to highlight the *Yishuv* successfully putting Zionist development into practice at the turn of the twentieth century.

Utopia

According to Fredric Jameson, the interpretation of utopia offers a *way of reading* which reinforces the 'political dimension' of a writer's text: 'by reading the very content and the formal impulse of the texts themselves as figures – whether of psychic wholeness, of freedom, or of the drive toward Utopian transfiguration – of the irrepressible revolutionary wish' (1972, 159). Jameson highlights that utopias express not an expectation for the future but rather provide a radical recreation of the present. In 'The politics of utopia' Jameson reaffirms that utopia enables the imagination of alternative political possibilities by creating 'the conception of systemic otherness, of an alternate society' (2004, 36) and, thus the political dimension is inherent in the way utopia enables the writer to critique the existing society and propose a new one. Paul Ricœur also stresses the potential for critique in utopia:

[at] a time when everything is blocked by systems which have failed but which cannot be beaten...utopia is our resource. It may be an escape, but is also the arm of critique. It may be that particular times call for utopia. (1986, 300)

Nevertheless, a common criticism of the use of utopia is that the literary imaginaries that writers produce may only have limited transformative power of the world *outside* of the text. Specifically, the narrative does not concern itself with how to change existing circumstances in order to create this new society because it derives its vision of an already transformed society from the writer's imagination.

However, Herzl was determined that his use of the form of utopia should not be conceived of in these terms. Herzl's novel departs from the generic characteristics of utopian fiction which frequently make overt or covert reference to previous utopias. One such example is William Morris' (1941) *News From Nowhere*, which responds directly to Edward Bellamy's (1980) *Looking Backward: 2000–1887* in which the protagonists of both novels fall asleep and wake up several centuries later to discover the state structure and ideas about labour and property have been transformed. Indeed, Herzl distanced his work from the fantastical or the idea that simply the influence of his own vision was sufficient. The note that accompanied the copy of *Altneuland* that he gave to Lord Rothschild[4] read: 'There will, of course, be stupid people who, because I have chosen the *form* of a Utopia...will declare the *cause* to be a Utopia. I fear no such misunderstanding in your case' (1960, 1357). Herzl's conception of the *cause* coincides with Ernst Bloch's view about the function of utopia. According to Bloch, utopia articulates 'anticipatory consciousness' which is an awareness of alternatives that have yet to take shape in the world *outside* the text but that could one day be put into practice. He demonstrates how art and culture can be utilised to create hope and a blueprint for an alternative future (1986). Herzl's assertion about 'the *cause*' or achievability is reinforced by the narrative which does not simply represent the creation of a new, alternative society. Rather, Herzl depicts how the Zionist discourses of nation-building and development will be put into practice in order to create such a society. The epigraph of Aldous Huxley's *Brave New World* by Nicolas Berdiaeff expresses a warning about the construction of utopias:

Utopias appear to be a good deal more realizable than was previously thought. And today we are faced with an alarming question of a different nature: How to avoid their complete realization? Utopias are realizable. Life moves towards utopias. And perhaps a new century is beginning, a century when intellectuals and the cultured class will dream of ways of avoiding utopias and of returning to a non-utopic society, less 'perfect' and more free. (2007, xxxix)

In other words, rather than fearing that utopia will never be realised, there is a fear that it will. Critics of utopianism have often asserted that its chief danger is its uncompromising, totalising character (Jameson 1991, 41). As I shall show in more detail below, Herzl's utopian Zionism as represented in the novel often conflicts with the geopolitical utopia of early Palestine.

According to Gabriel Piterberg, Herzl's novel is not simply utopian; rather 'it is a utopian *colonial* novel' (2008, 37). Piterberg's use of 'colonial' to describe the characteristics inherent in Jewish immigration to and creation of a national home in Palestine is very controversial in Israeli academia. Zeev Sternhell critically engages with Piterberg's study and challenges the account of Jewish nationalism's origins and trajectory. Sternhell argues that 'Zionism was a stringent nationalism, a radical nationalism; but to claim that the arrivals were white settlers driven by a colonialist mind-set does not correspond to historical reality' (2010, 111). With the exception of revisionist historians including

Gershon Shafir, Ilan Pappé and Baruch Kimmerling, much of Israeli academia vehemently denies Zionism can be considered a form of colonisation because the Jewish conquest of land and labour was not exploitative of the indigenous population but rather based on nationalism and the creation of a Jewish national home. Piterberg's observation of the 'colonial' characteristics of the alternative society Herzl envisions is important because, unlike previous critics, he refers to 'utopia's spatial consequences' (2008, 37). However, despite this, Piterberg refrains from elaborating further about how Herzl's novel depicts land use and agriculture and Zionist collectivism, in other words the material means by which colonisation is achieved in this novel.

The novel's narrative trajectory that traces the transformation of the social order reinforces Herzl's use of utopia. As Peter Fitting observes, 'most literary utopias in fact juxtapose a critique of the author's existing society with the description of the better society...the proportion of social critique and the positive description of a new social system can vary from work to work' (2010, 141). In the first book of *Altneuland*, the narrative set in Palestine depicts the poverty associated with underdevelopment and a dispersed Jewish community. The second and third books represent prosperity through agricultural development and a collective Jewish community.

Altneuland

In the opening pages of the novel, Friedrich is presented as depressed and isolated from the community in Vienna in which he lives. He is 'in deep melancholy' and '[h]e felt too tired to make new acquaintances' (3). Friedrich's separateness from society resonates with the historical context and the Jewish Question of assimilation versus segregation in Europe. In his essay 'On the Jewish Question', Karl Marx regards the primary responsibility for Jewish equality and emancipation in Europe as residing with Jewish communities because they reject the opportunity for assimilation into a society that was responsible for Jewish segregation. Segregation took the form of the ghetto in Eastern Europe and social alienation in Western Europe (1978, 26). Thus, Herzl's characterisation of Friedrich reflects this social alienation and also links with his use of the *Bildungsroman* form which provides a teleology from exclusion into integration in the narrative. The narrative trajectory traces Friedrich's development from a marginal position as a Jew in Europe towards an integrated individual in society, thereby highlighting the 'demarginalising' potential of the *Bildungsroman* (Slaughter 2007, 134). It is precisely Friedrich's social alienation that makes him not only the typical *Bildungsheld* but also the ideal candidate to be Kingscourt's travelling companion:

> I must remind you that you are undertaking a life-long obligation...If you come with me now, there will be no going back. You must cut all your ties.
>
> 'Nothing binds me,' replied Friedrich. 'I am all alone in the world, and have had enough of life.' (22–23)

On the course of their voyage, Friedrich and Kingscourt make their first visit to Palestine. On disembarking in Jaffa, they survey the landscape:

> Jaffa made a very unpleasant impression upon them. Though nobly situated on the blue Mediterranean, the town was in a state of extreme decay. Landing was difficult in the forsaken harbor. The alleys were dirty, neglected, full of vile odors. Everywhere misery in

bright Oriental rags. Poor Turks, dirty Arabs, timid Jews lounged about-indolent, beggarly, hopeless. A peculiar, tomblike odor of mold caught one's breath. (29)

The disgusted tone combined with the repetition of 'odour' and references to 'decay', 'dirt' and 'mould' reinforces a sense of repulsion with their surroundings. The reference to inhabitants subverts the 'myth of Jewish ignorance of the presence of Palestinian Arabs' (Shafir 1996, 204) as epitomised in the slogan 'a land without people, for a people without a land'.

Likewise in Jerusalem, Friedrich and Kingscourt encounter:

a picture of desolation. The lowlands were mostly sand and swamp, the lean fields looked as though burnt over. The inhabitants of the blackish Arab villages looked like brigands. Naked children played in the dirty alleys…The bare slopes and the bleak, rocky valleys showed few traces of present or former cultivation. (30)

The overlap in Herzl's diaries and his novel can be seen through the passages from his diaries that correspond to the narrative of *Altneuland*. His description of Jerusalem in his diaries: 'The musty deposits of two thousand years of inhumanity, intolerance, and uncleanliness lie in the foul-smelling alleys' (1960, 745) highlights the direct echoes between the two, specifically the use of the same language and disgusted tone. The word choice of 'brigands' suggests that the inhabitants are perceived as outlaws in a wild or isolated terrain. This view also ties in with Herzl's assertion in *A Jewish State* about the Jewish need create a state that would provide 'an outpost of civilisation as opposed to barbarism' (1896, 30). Herzl's representation of the landscape reflects the intimate connection between travel writing and colonialist discourse that existed in the nineteenth century. This representation was aided by many early twentieth-century European travel accounts of Palestine which depicted the land as poor and neglected and the population as impoverished or weak. Mark Twain writes about how he discovered on his travels: the 'desolate country whose soil is rich enough, but is given over wholly to weeds… a silent mournful expanse…a desolation' (1869, 488). In a German encyclopaedia published in 1827, Palestine was depicted as 'desolate and roamed through by Arab bands of robbers' (Brockhaus 1827, 206). Mary Louise Pratt argues that landscape in literature and art refers to the particular gaze or perception of 'a European male subject of European landscape discourse – he whose imperial eyes look out and possess' (1992, 7). As such, Herzl's/ Friedrich's gaze is one of control and projects a fabricated set of Eurocentric preconceptions and authoritative vision. In such an Orientalist paradigm, the textual representations of indigenous peoples are depicted as Other (Said 2003). Derek Penslar argues that 'Zionist thinking, like that of *fin de siècle* Europeans as a whole, operated on multiple levels and that feelings of benevolence, humanitarianism, and sympathy could easily blend with condescending, Orientalist, and even racist views of the Palestinian Arabs' (2003, 84). The significance of a socially alienated European Jew seeing Palestine as a desolate land and its native inhabitants feckless is that this image was reflected in paintings and photographs from this period and helped bolster the need for the Jewish immigrants to 'tame the wilderness' and 'make the desert bloom'.[5]

The passage about Jaffa also demonstrates that the poverty and deprivation of the Arabs are perceived to be directly correlated to their inability or refusal to work the land. Specifically, presenting the land as 'a picture of desolation and neglect' suggests the lack of cultivation and production prior to Jewish settlement. Sandra Sufian recognises that

[u]nder the *terra nullius* principle, applied particularly in colonial contexts, if the land was not being cultivated, then by Western standards it was considered as not being properly used. Those who could, therefore, cultivate the land had the right, if not an obligation, to do so. (2007, 46)

It was precisely this perceived failure to treat the land with the proper or required care and attention that Zionism used to legitimate the Jews' claim to the land and concomitantly the Arabs' forfeiture of this land. The description of the early period of Zionist settlement by Shimon Peres, former Prime Minister of Israel, reflects this attitude:

[t]he land to which they [the Jewish settlers] came, while indeed the Holy Land, was desolate and uninviting; a land that had been laid to waste, thirsty for water, filled with swamps and malaria, lacking in natural resources. And in the land itself there lived another people; a people who neglected the land, but who lived on it. (cited in Said and Hitchens 2001, 5)

This view of the Palestinian inhabitants and the land formed a key component of the Zionist colonial project and nation-building. The land of Palestine was frequently perceived and described as a locus of contaminated conditions of polluted water, swamps and contagious diseases such as malaria (Sufian 2007). Moreover, these two passages demonstrate that the poverty and disease used to characterise the Arabs' underdevelopment are perceived to be an inherent characteristic of their culture. However, Alon Tal contradicts the depiction of Palestine as a wasteland in his environmental history of Israel:

The question of how barren the land of Israel was prior to Zionist settlement has been highly politicised in the 'tit-for-tat' debates between pro-Israeli and pro-Arab camps. Environmentalists now question the appropriateness of such loaded terms as 'barren' and 'desolate,' given the remarkable underlying biodiversity. (2002, 37)

The perception of Arab underdevelopment can also be countered by the fact that in Palestine, as early as 1800, the population was an estimated 250,000–300,000 inhabitants, the 'vast majority lived as peasants' and 'agriculture formed the basis of the local economy and society' (Kramer 2008, 44–45). As such, the land was cultivated for subsistence crops and trade.

Friedrich explains how he perceives the poor and difficult circumstances of the Jewish people to be reflected in the condition of the homeland: "'If this is our homeland," remarked Friedrich sadly, "it has declined like our people"' (30). The concept of the Jewish people being inextricably linked with the land of Israel is prominent in Zionist discourse. As Yael Zerubavel argues: 'Zionism assumed that an inherent bond between the Jewish people and their ancient land was a necessary condition for the development of Jewish nationhood' (1994, 15). Kingscourt adheres to this explicit connection:

'Yes, it's pretty bad,' agreed Kingscourt. 'But much could be done here with afforestation, if half a million young giant cedars were planted – they shoot up like asparagus. This country needs nothing but water and shade to have a very great future.' 'And who is to bring water and shade here?' 'The Jews!' And Kingscourt swore a great cavalry oath. (30)

Whilst the text represents the characters imagining an unrealised future, the significance of land use in this passage is that the planting of trees demonstrates Herzl's legacy and influence on the material environment of Israel/Palestine. From the beginning of the twentieth century, tree planting has functioned to assert and naturalise the Jewish settler's presence on the land whilst concomitantly submerging the Palestinian presence (Boast

2012: 47). Additionally, the planting of trees and other agricultural methods were recognized by the Yishuv as key ways to 'redeem' the land. Herzl is evidently prefiguring the idea of Zionist environment discourse which Shaul Cohen defines in the following terms:

> the environment has been the stage upon which the Zionist enterprise has been built, and its features are opportunities to showcase good stewardship, and the prowess of Jewish agriculture and development...In Zionist ideology, redemption of the self comes through redemption, i.e. rehabilitation of the land. (2011, 248)

As such, the idea of 'redemption' becomes a means of regaining the land – if we work the land, we will not lose it again. Working the land also linked to the idea of culpability as through 'redeeming' which equates to farming, the Jewish settlers perceive that they are regaining lost land and thereby avoid losing the land again in the future. Also, the Jewish community would undergo a radical transformation through their participation in physical labour and 'redeem' not only the land but also the whole community.

Moreover, when the two men return to Palestine after a period of two decades, they discover that through Jewish immigration and labouring on the land, there has been a socio-economic transformation of the region. "'How changed it all is!' cried Friedrich. 'There's been a miracle here'" (40). The representation of the landscape provides a striking juxtaposition with the passages on Jaffa and Jerusalem cited above:

> On both sides of the road there were carefully tended fields, vineyards, tobacco plantations, tree nurseries, with nowhere so much as an inch of wasteland. Some way off they saw a machine mowing a field of clover. Now and again a cart passed them high with Lucerne for fodder. Here and there the summer crops had already come up – maize and sesame, lentils and vetch. On the fallow land motor-ploughs were preparing the fields, still moist from the spring rains, for the next sowing. (120–121)

The differences between the two depictions lie not only in the fertility and irrigation that is evident in the latter passage but also the role of agricultural technology in enabling this change to take place, both characteristics that were absent in the first passage. Agricultural technology is also given a prominent place through the enumeration of 'a machine', 'a cart' and 'motor-ploughs' (120–121). Alon Tal asserts that

> many Zionist leaders like Herzl can be readily diagnosed as 'technological utopians'. For them, technology, if pursued with sufficient ardour, constituted the most reliable instrument for attaining power over their surroundings and for overcoming the menacing and primitive 'wastelands'. (2008, 281)

Consequently, labour and technology are projected as the necessary components to establish a prosperous and successful national enterprise based on agricultural production. As Glenna Anton acknowledges, Zionism from the outset 'celebrated technological innovation and modernization while at the same time idealized a close connection to the soil through labour' (2008, 90). There is little tension between technological and agrarian utopianism. As such, the transformation from what was perceived to be a traditional unproductive form of agriculture to modern technological agriculture is depicted as crucial to the success of settler-colonialism. In the initial stages of Jewish settlement in Palestine prior to World War I, the use of technological agriculture to redeem the land was to be realised through the use of synthetic fertilisers, However, Alon Tal argues that fertilisers have caused serious environmental damage:

> Zionism spawned a high input, 'technological' agriculture. For instance, a key to the successful land reclamation by Jewish farmers was synthetic fertilizers…Years later the nitrates would reappear as high concentrations…[thereby creating] a hydrological hazard. (2007, 5–6)

In a further article, Tal refers to need for the emergence of a 'pro-environmental, technologically sceptical' attitude because 'technology has led Israel into its present environmental crisis' (2008, 302). As such, Jewish efforts to 'make the desert bloom' involved cultivation and fertilisation methods that proved to have negative repercussions on the environment today.

Furthermore, in *Altneuland* Herzl projects a reiteration of the idea of the redemption of the land being directly correlated to the redemption of the Jewish people:

> [w]e took our children out of damp cellars and hovels, and brought them into the sunlight. Plants cannot thrive without sun. No more can human beings. Plants can be saved by transplantation into congenial soil. Human beings as well. That is how it happened. (55)

This dual redemption is emphasised in Zionist discourse, specifically Max Nordau's, a close acquaintance of Herzl's, vision of 'Judaism with muscles' (1980, 434–435) which was needed in order 'to replace the pale-faced and thin-chested 'coffee-house Jew', and to regain the heroism of his forefathers in the land of Zion' (Weiss 2002, 1). In other words, Jewish settlement and agricultural labour in Palestine regenerated Jewish society and the land. This idea is more thoroughly explained to Friedrich, Kingscourt and the reader:

> The wealth of a country lies in its workers…The more workers that come along, the more bread there is – provided the society is as just as ours…just as it is good for Newville if it grows, more and more settlements being added to its outskirts, so it is good for the New Society…The more men come here to work, the better off we shall be…The eldest among you know what this village was like twenty years ago, all empty and desolate…Stones were cleared, swamps drained…And today Newville is a garden, a lovely garden where you live a good life. (118)

Herzl depicts how the workers put Zionist discourse into practice through settlement and the labour of the Jewish workers. Kramer argues that the aim of 'the idea of "productivizing" the "Jewish masses," by transforming Jewish *Luftmenschen*[6] into workers… was to establish an egalitarian Jewish society that would be largely self-contained and self-sufficient' (2008, 111). Zionist environmental practice and land use is also represented as the initial stage of this project: 'Stones were cleared, swamps drained'. The transition into a garden resonates with Herzl's description in *The Jewish State* about how Zionism is able to bring technology to Palestine: '[w]here we modern ones appear with our inventions we transform the desert into a garden' (1896, 99). As we have seen, Zionist rhetoric and the Jewish immigrants of *Altneuland* sought to 'redeem' and assert a claim to the land through an ability to 'make the desert bloom' with modern technology that presented Zionist settlers as more 'civilised' than the Palestinian farmers.

The novel illustrates that the combination of settlement and agricultural development is not only exclusively beneficial to the Jewish community in Palestine but also produces advantageous effects for the Palestinian Arab community. The sole Arab character of the narrative is Rashid Bey, who appears to act as a mouthpiece for the effects of the Zionist project on his community, is depicted as being 'one of those who immediately grasped that Jewish immigration could only be beneficial to all, and he profited from our economic boom' (54). As Bey explains about Jewish immigration himself:

> It was a great blessing for all of us. …those who had nothing stood to lose nothing, and could only gain. And they did gain: opportunities to work, prosperity. Nothing could have been more wretched than an Arab village at the end of the nineteenth century. The peasants' clay hovels were unfit for stables. The children lay naked and neglected, and grew up like dumb beasts. Now everything is different. They benefited from the progressive measures of the New Society whether they joined it or not. The swamps were drained, the canals built…the natives were the first to be employed, and were well paid for their work! These people are better off than at any time in the past. (81–83)

Bey's representation of the land prior to Jewish settlement as uncultivated and the Arab population as impoverished and neglectful of the land echoes the Jewish depiction above. Earlier in the same scene, Kingscourt and Friedrich are told by one of their guides, 'Do not expect to see the filthy nests that used to be called villages in Palestine' (81). As such, the landscape has been stripped off the presence of indigenous Arab villages. Bey also emphasises prosperity and better standards of living for the Arab inhabitants of Palestine. Indeed, Israel's advocates continue to assert that living standards and political freedoms of Israeli Arabs are superior to Arabs living in surrounding countries. Shlomo Avineri regards such ideas in the novel as evidence of Herzl's 'tolerance and universalistic humanitarianism, characteristic of his Central European outlook and his impeccable vision of civil rights as related to the Palestinian Arabs' (1981, 99). However, in comparison to the Jewish community, the Arab inhabitants of Palestine in the period when *Altnueland* is set, experienced rising inequity in political power and economic conditions (Pappé 2011, 110). Ron Smith recognises that Zionist discourse like other forms of colonial discourse embodies: '[t]he binary of utopia/dystopia' (2012, 21) for Jews and Palestinians in Israel/ Palestine. Indeed, this binary is evidenced in the increasing material inequity between the two populations in Israel/Palestine today. During the period in which the novel is set, Labour Zionists placed significant ideological importance on 'conquest of labour' which related to the first Jewish settlers competing with the indigenous Arab population for labouring positions in the agricultural colonies (Shindler 2008, 18–20). However, this struggle proved futile as Arab labourers continued to be employed over Jewish labourers in the Jewish sector of the Jewish economy of Palestine. The solution to scarce labouring positions and employment was the conceptualisation of 'Hebrew labour' which excluded the indigenous labourers and led to the establishment of agricultural colonies that were separate from Arab villages.

The utopian narrative of *Altnueland* illuminates Herzl's vision of the move away from a perceived wasteland with feckless inhabitants towards a cultivated land and a mutually beneficial society for both Jews and Arabs – but only on the terms dictated by Zionist development discourse. Postcolonial ecocritics need to become attuned to how Herzl's depiction of the Zionist settler-colonial project offers an *unrealised* imagining of the environmental and social transformation of Mandate Palestine. The need for such recognition is all the more critical, given the fact that *Altneuland* 'is still an accurate portrayal of the way the majority of Israelis (and Diaspora Jews) understand the unfolding of the Zionist-sponsored "development" of Palestine' (Levine 2005, 182). Consequently, it is clear that *Altneuland* continues to function as a source of the enactment and assertion of

Zionist development because Herzl's fictional representation of the simultaneous redemption of the land and Yishuv is a recurring one and still being unproblematically accepted and disseminated today. The narrative closure of utopia that Herzl produces demonstrates his enduring legacy of creating the narrative genre of successful environmental transformation. Examining both geopolitical and the literary utopian projects is all the more essential given that the Zionist utopian project has created an ongoing dystopia for Palestinians over a century after it was first conceived.

Notes

1. My reasoning for the inclusion of Palestine in postcolonial studies is derived from the work of Anna Bernard. Bernard argues that 'If you take the field of "postcolonial studies" to describe the study of European high imperialism and its aftermath, as I do, then it also includes…the British Mandate' (2010, 1). See also Bernard (2013). As such, the fact that Herzl wrote and set part of *Altneuland* during the period when Mandate Palestine was under British colonial rule means that the text can be considered within the rubric of postcolonial studies.
2. Measures which include the creation of agricultural settlements and cultivation that were taken in order to gain to control of land and, thus exercise territoriality are illustrated by the term 'colonisation'.
3. *Yishuv* is a Hebrew term designating the social structure of the Jewish community in Palestine from the Second Aliyah period (1903–1914) until 1948 and the establishment of the state of Israel. See Kramer (2008, 101).
4. The financial support of the Rothschilds helped the Yishuv to settle in Palestine by providing the capital to purchase land.
5. The creation of agrarian settlements was inspired by Labour Zionism and 'making the desert bloom' was the vision of David Ben-Guiron, the future first Prime Minister of the state of Israel. This rhetoric also resembles that of pioneers of the American West: a land represented as 'lacking' the settlers' transformations thanks to 'incompetent and irresponsible' native inhabitants.
6. The Yiddish term *Luftmenschen* refers to Jewish people 'inclined to speculative, economically unproductive occupations' (Miron 2000, 24).

References

Anderson, B. 1991. *Imagined Communities: Reflections on the Origin and Spread of Nationalism*. London: Verso.

Anton, G. 2008. "Blind Modernism and Zionist Waterscape: The Huleh Drainage Project." *Jerusalem Quarterly* 9 (1): 76–92.

Avineri, S. 1981. *The Making of Modern Zionism: Intellectual Origins of the Jewish State*. London: Weidenfeld & Nicolson.

Bar-Yosef, E. 2007. "A Villa in the Jungle: Herzl, Zionist Culture, and the Great African Adventure." In *Theodor Herzl: From Europe to Zion*, edited by M. H. Gelber and V. Liska, 85–102. Tubingen: Max Niemeyer.

Bellamy, E. 1980. *Looking Backward: 2000–1887*. London: George Routledge and Sons.

Berdiaeff, N. 2007. "Epigraph." In *Brave New World*, edited by Aldous Huxley, xxxiv. London: Vintage.

Bernard, A. 2013. Rhetorics of Belonging: Nation, Narration, and Israel/Palestine. Liverpool: Liverpool University Press.

Bernard, A. 2010. "Palestine and Postcolonial Studies." Paper presented at London Debates 2010, University of London, May 13–15. Accessed August 12, 2013. On line at http://events.sas.ac.uk/fileadmin/documents/postgraduate/Papers_London_Debates_2010/ Bernard__Palestine_and_postcolonial_studies.pdf

Bloch, E. 1986. *The Principle of Hope*. Translated by Neville Plaice, Stephen Plaice, and Paul Knight. Oxford: Blackwell.

Bloom, E. 2011. *Arthur Ruppin and the Production of Pre-Israeli Culture*. Boston: Brill.

Boast, Hannah. 2012. "'Planted Over The Past': Ideology and Ecology in Israel's National Eco-Imaginary.' *Green Letters: Studies in Ecocriticism* 16 (1): 46–58.

Brockhaus. 1827. *Alig. deutsch Real-Encyklopaedie*. 7th ed., vol. 8. Leipzig: Bibliographisches Institut.

Cohen, S. 2011. "Environmentalism Deferred: Nationalisms and Israeli/Palestinian Imaginaries." In *Environmental Imaginaries of the Middle East and North Africa: History, Policy, Power and Practice*, edited by D. K. Davis, and E. Burke, 246–264. Athens: Ohio University Press.

Fitting, P. 2010. "Utopia, Dystopia and Science Fiction." In *The Cambridge Companion to Utopian Literature*, edited by G. Claeys, 135–153. New York: Cambridge University Press. doi:10.1017/CCOL9780521886659.006.

Herzl, T. 1896. *A Jewish State: An Attempt at a Solution of the Jewish Question*. Translated by Sylvie D'Avigdor. London: D. Nutt.

Herzl, T. 1902. "Old-New Land." *The Maccabean* 3 (4).

Herzl, T. 1960. *Complete Diaries*. Translated by Harry Zohn. New York: Herzl Press and Thomas Yoseloff.

Herzl, T. 2011. *Old New Land*. Translated by Lotte Levensohn. North Charleston, SC: CreateSpace Independent Publishing Platform.

Huggan, G., and H. Tiffin. 2010. *Postcolonial Ecocriticism*. London: Routledge.

Jameson, F. 1972. *Marxism and Form: Twentieth-Century Dialectical Theories of Literature*. Princeton, NJ: Princeton University Press.

Jameson, F. 1991. *Postmodernism, or, the Cultural Logic of Late Capitalism*. London: Verso.

Jameson, F. 2004. "The Politics of Utopia." *New Left Review* 25: 35–54.

Kramer, G. 2008. *A History of Palestine: From the Ottoman Conquest to the Founding of the State of Israel*. Translated by Graham Harman, and Gudrun Kramer. Princeton, NJ: Princeton University Press.

Levine, M. 2005. *Overthrowing Geography: Jaffa, Tel Aviv, and the Struggle for Palestine, 1880–1948*. Berkeley: University of California Press.

Marx, K. 1978. "On the Jewish Question." In *The Marx–Engels Reader*, edited by R. C. Tucker, 26–52. New York: Norton.

Masalha, N. 2007. *The Bible & Zionism: Invented Traditions, Archaeology and Post-Colonialism in Palestine–Israel*. London: Zed Books.

Miron, D. 2000. *The Image of the Shtetl and Other Studies of Modern Jewish Literary Imagination*. Syracuse, NY: Syracuse University Press.

Morris, W. 1941. *News from Nowhere*. London: Nelson.

Nordau, M. 1980. "Jewry of Muscle." In *The Jew in the Modern World: A Documentary History*, edited by P. Mendes-Flohr, and J. Reinharz, 434–435. Oxford: Oxford University Press.

Pappé, I. 2011. *The Forgotten Palestinians: A History of the Palestinians in Israel*. New Haven, CT: Yale University Press.

Penslar, D. J. 2001. "Zionism, Colonialism and Postcolonialism." *Journal of Israeli History: Politics, Society, Culture* 20 (2–3): 84–98. doi:10.1080/13531040108576161.

Piterberg, G. 2008. *The Returns of Zionism: Myths, Politics and Scholarship in Israel*. London: Verso.

Pratt, M. L. 1992. *Imperial Eyes: Travel Writing and Transculturation*. London: Routledge.

Ricœur, P. 1986. *Lectures on Ideology and Utopia*. New York: Columbia University Press.

Ruppin, A. 1976. *The Agricultural Colonisation of the Zionist Organisation in Palestine*. Translated by R. J. Feiwel. Westport, CT: Hyperion Press.

Said, E. 2003. *Orientalism*. London: Penguin Books.

Said, E. W., and H. Christopher. 2001. *Blaming the Victims: Spurious Scholarship and the Palestinian Question*. London: Verso.

Shafir, G. 1996. *Land, Labor, and the Origins of the Israeli-Palestinian Conflict, 1882–1914*. Berkeley: University of California Press.

Shindler, C. 2008. *A History of Modern Israel*. Cambridge: Cambridge University Press.

Slaughter, J. 2007. *Human Rights, Inc.: The World Novel, Narrative Form, and International Law*. New York: Fordham University Press.

Smith, R. J. 2012. "Geographies of Dis/Topia in the Nation-State: Israel, Palestine, and the Geographies of Liberation." *TDSR* 23 (2): 19–33.

Sternhell, Z. 2010. "In Defence of Liberal Zionism (a Review of Piterberg's Returns of Zionism)." *New Left Review* 62: 99–114.

Sufian, S. M. 2007. *Healing the Land and the Nation: Malaria and the Zionist Project in Palestine*. Chicago, IL: University of Chicago Press. doi:10.7208/chicago/9780226779386.001.0001.

Tal, A. 2002. *Pollution in a Promised Land*. Berkeley: University of California Press.

Tal, A. 2007. "To Make a Desert Bloom: Seeking Sustainability for the Israeli Agricultural Adventure." Accessed August 22, 2013. http://www.yale.edu/agrarianstudies/colloqpapers/01tal.pdf

Tal, A. 2008. "Enduring Technological Optimism: Zionism's Environmental Ethic and Its Influence on Israel's Environmental History." *Environmental History* 13: 275–305. doi:10.1093/envhis/13.2.275.

Twain, M. 1869. *The Innocents Abroad*. Hartford, CT: American Publishing Company.

Vital, A. 2008. "Toward an African Ecocriticism: Postcolonialism, Ecology and *Life & Times of Michael K*." *Research in African Literatures* 39 (1): 87–106. doi:10.2979/RAL.2008.39.1.87.

Weiss, M. 2002. *The Chosen Body: The Politics of the Body in Israeli Society*. Stanford, CA: Stanford University Press.

Zerubavel, Y. 1994. *Recovered Roots: Collective Memory and the Making of Israeli National Tradition*. Chicago, IL: University of Chicago Press.

'The Republic of Arborea': trees and the perfect society

Shelley Saguaro

School of Humanities, University of Gloucestershire, Gloucestershire, UK

This essay explores the persistence of trees or forests as a utopian locus of liberty and individual authenticity, with reference to two texts, Italo Calvino's *The Baron in the Trees* and Sam Taylor's *The Republic of Trees*. Both texts have some commons themes, such as the adolescents' escape from authority by retreating to the forest. The young revolutionaries in each text are most at home with/in trees; both also engage explicitly with Rousseau's revolutionary idealism. Calvino's fable, set in the eighteenth century, recounts the tale of a young aristocrat, Cosimo, who at age 12, rebels against his family and renounces forever any actual contact with the earth. Cosimo's younger brother, the narrator, ends the novel with the grim acknowledgement of the changes about to be wrought in the new century, the nineteenth, where trees themselves 'seem almost to have no right here', 'swept away by this frenzy for the ax', and where 'old' and 'natural' indigenous trees are displaced by imports from 'Africa, Australia and the Americas, the Indies' (217). *The Republic of Trees* is also about a young rebel who, with his companions, flees from authority and takes up residence in the forest, but the context is the twenty-first century. The four friends who have run away to live in the woods *take A History of the French Revolution* and Rousseau's *The Social Contract* to aid them, but in a tale that is also reminiscent of *The Lord of the Flies*, the consequences are disastrous. The article discusses the politics of utopia and dystopia, in both an Enlightenment and post-Enlightenment context, including the way that these ideologies impacted on forests and on bio-relations. Finally, the article proposes that the declaration of the rights of trees and all nature, as outlined in Cosimo's 'dead letter' treatise, may now have found its viability in a new forest locus and political context: Ecuador.

There is in European and western literature a persistence of reference to trees or forests as a utopian locus of liberty and individual authenticity. Who can conceive of a Utopia without trees? From Eden's Tree of Knowledge to the Liberty Trees of the American and French revolutions, the tree has a symbolic aspect which, in its 'cultural surplus', transcending cultural and historical specificities, is a 'prefiguration of wholeness or a better way of being' (Levitas 2013, 5). Collectively, as forest, the wild wood is also to be greatly feared, as many an instructive fairy tale or illustrative allegory will attest. Although forests have often constituted an atmospheric backdrop in literature, one way or the other, they have lately also been a topic for scientists, environmentalists and politicians. This is for various reasons but above all, it is because of their general and often drastic diminishment across the globe. Recently, however, and this is also in the contemporary context of ecological crisis, trees have been discussed with a new emphasis.

They are construed, in a democratic and utopian vein, as being integral to a good society, not simply as 'the backdrop for human action, a set of resources, or another set of interests' (Sargisson 2013, 128) but, rather, as citizens and subjects with whom humans interact, to the benefit and flourishing of both. The dominion that mankind has assumed over the natural world has become indefensible; as Marius de Geus noted as early as 1999 in *Ecological Utopias: Envisioning the Sustainable Society*: 'an ecologically viable society' requires not just 'an awareness of environmental issues'; what is required in no less than 'a completely different attitude towards nature' (cited in Sargisson 2013, 117). Bruno Latour in *Politics of Nature: How To Bring Sciences Into Democracy*, reinforces the view, along with many others (Andrew Dobson, John S. Dryzek, Daniel Botkin, Timo Maran, Timothy Morton and Jane Bennett, to name a few), that: 'nothing is more anthropocentric than the inanimism of nature' (Latour 2004, 224). He adds his own call for change: 'I am asking [….] that the question of democracy be extended to nonhumans' (Latour 2004, 223). Jane Bennett goes even further, developing a 'Creed' that states: 'I believe it is wrong to deny vitality to non-human bodies, forces and forms [...] even though it resists full translation and exceeds my comprehensive grasp' (Bennett 2010, 122). Recent utopian writing also calls for the 'rewilding' (Monbiot 2013) of an over-civilised world which 'limits the range of our engagement with nature' and 'pushes us towards a monoculture of the spirit' (Monbiot 2013, 154). 'Of all the world's creatures,' states George Monbiot in his recent *Feral: Searching for enchantment on the frontiers of rewilding*, 'perhaps those in greatest need of rewilding are our children. The collapse of children's engagement with nature has been even faster than the collapse of the natural world' (Monbiot 2013, 167).

Two literary texts, one written in the twentieth century, Italo Calvino's *The Baron in the Trees* (1957) and the other in the early twenty-first, Sam Taylor's *The Republic of Trees* (2005), while not programmatically utopian or specifically ecotopian, are considered here as writing which inspires us to think differently about the societies in which we live, and both anticipate or put to work views currently being developed as topical and crucial. Both focus on human relation to and relationship with trees; the liminality of adolescence is used to show both the paucity of cultural conventions and utopian potential. Both narrate 'rewilding' with variable outcomes. In particular, they are both instructive and thought-provoking in terms of rethinking 'nature'. Utopianism is also integral to both, not least in their shared reference points to the prior – and to a large extent drastic – utopian tenets and projects of the French Revolution. The new Republics posited by both these texts, albeit fancifully or obliquely, reprise an interrelationship of humans with trees – and more.

Written and published 50 years apart, both *The Baron in the Trees* and *The Republic of Trees* have a clear titular similarity, and some common themes, such as the adolescents' escape from authority or restriction by retreating to the freedom of the forest. Italo Calvino (1923–1985) had a long and prolific writing career which started in Italy while he was a member of the Communist Party. However, the apparently fantastical tale *The Baron in the Trees*, which was published in the same year Calvino resigned from the Party, was not a departure from politics. As Calvino explained in a letter to his colleague and friend, Paolo Spriano, in August 1957, resigning from an institution did not mean resigning from one's clear utopian aspirations:

> It is difficult being a Communist on your own. But I am and remain a Communist. If I can manage to prove that to you, I will also have proven that *Il barone pampante* (*The Baron in the Trees*) is a book that is not too far from the things that we are really interested in. (Calvino 2013, 139)

Sam Taylor, whose literary career is relatively new with three novels published to date, cites Calvino as a writer who has inspired his own work (in Cornwell 2005); although not specifically stated, there are many indications that *The Republic of Trees* is an allusive and intertextual engagement with its precursor, *The Baron in the Trees*. In a podcast on his website, Sam Taylor explains that the novel, with its four adolescents who 'escape to the forest', is about 'the effects of first love': 'jealousy, memory, heartbreak', but that it is also about 'the forest'. He refers to the importance of a memory used in the novel: his own 'dull' existence growing up in suburban housing-state in 1970s England, when a distant view of forest was 'the only object on the horizon that was mysterious in any way' (http://www.samtaylor.com/books.htm#republic_reviews).

The other feature that these texts have in common is the eighteenth century, and in particular, the utopian and republican, and indeed, arboreal, ideals of Jean Jacques Rousseau. The young revolutionaries in each text, one an eighteenth-century Italian aristocrat, Cosimo di Rondò, and the other, Michael, a twenty-first century orphan who moves from English suburbia to rural Provence, are most at home and feel most them-selves with/in trees. Both texts, with their diverse settings, also engage explicitly with Rousseau's revolutionary idealism and his theories on recuperating 'natural man'; Cosimo is Rousseau's contemporary, and he pens treatises in the manner of, and sometimes for the attention of, the Encyclopaedists and *philosophes*, Rousseau, Diderot, Voltaire and Montesquieu, one such example being '*Project for the Constitution of an Ideal State in the Trees*, in which he described the imaginary Republic of Arborea' (142). Michael and his comrades, on the other hand, establish their Republic of Trees inspired by Rousseau's *The Social Contract* and stimulated by reading *The History of the French Revolution*. Calvino's text is a parable or fantastic fable where the eighteenth-century protagonist lives a singular but rich life living in trees, ('something entirely good') (Calvino 1959, 213), and Taylor's a nightmarish post-Freudian narrative of adolescent angst, where doing what is good is utterly abandoned despite the 'purity of the trees' (Taylor 2005c, 179). Also common to both tales, despite the impossible fantasy of one or the dystopian horror of the other, is the social necessity of trees; trees matter for human well-being and mental equilibrium. Further, the active roles of trees, and thus their own *rights* as subjects of a utopian republic, are also suggested by both. Cosimo explicitly include the rights of trees, as other life forms, in his *Constitutional Project for a Republican City with a Declaration of the Rights of Men, Women, Children, Domestic and Wild Animals, Including Birds, Fishes and Insects, and All Vegetations, whether Trees, Vegetables, or Grass* (205). In the newly designated 'perfect society' (29) of the Republic of Trees, it is only Michael who views them as more-than-symbolic: 'I looked at the trees not as landmarks or symbols but as trees. They were my friends and this was my home' (Taylor 2005c, 220). His, albeit distorted, utopian vision includes the liberation of trees, mountains and birds in a life where people (and his own emotions) have let him down. As Michael replaces the usual filial 'three little words' ('I love you') for 'kill him now', the adolescent utopian enterprise ends in psychosis and murder and, as he stands momentarily poised on the limb of a tree viewing the beheaded victims below, a final loss of balance. Unlike *Lord of the Flies*, for instance, with which *The Republic of Trees* is often compared, the point of the tale is less the dystopia (and failure) of the adult-free wild than the dystopia that constitutes the wider familial and cultural norm that is their upbringing. Thus, the terrible failure of the republic in the forest is, particularly vividly in Michael's case, a result of the damage inflicted upon him by social and familial circumstances: (dead father, suicide-mother, abusive aunt) and the dislocated environments (English commercial suburbia; French utilitarian rural; at-distance forests) experienced in his childhood.

On trees

Both protagonists find they have a special relation to trees: a profound appreciation *of* them and a facility for being *in* them. For Michael Vignal, first from a suburban housing estate in the English Midlands and later from a Southern French farm village, 'basic-looking and long past its peak' (8), looking at distant trees or forests has fostered 'longing', awe and dreams of escape. Escaping to the forest ('we entered the forest and the forest entered our lives') (19), Michael finds different perspectives engender different feelings:

> I saw it, suddenly there, in front of us, below us and above us [...] the grand, dark sweep of the forest. It looked like an inky mirage, a black hole, a magic doorway through which we might disappear [...] Mostly what I remember is the fear [...] each time I looked up at the thin crack of sky bitten into by dark branches, I had to fight back a tide of panic. [...] Here there were no walls. Everything was alive. I sensed that the forest had eyes, that it was watching me. (18–19)

When Michael climbs up and swings through the trees, however, he finds that he is, though not unafraid, 'a natural', 'just like a little monkey' (28). He comes to believe, too, as he takes risks and leaps of faith that the trees save him: 'the tree caught me. The trees were my friends' (25). Up in the trees, Michael is 'instantly lighter' (41), free from thinking and fretting, but as soon as he touches ground again 'gravity did its work' (92) and he is consumed again by sexual jealousy and hallucinatory fear. Nature, in its purity, becomes his consoler and moral mentor: 'Nature helped. I thought of the trees' disappointment in me the day of my hangover' (179). Michael's understanding of, and relation to, nature in this precipitate Republic are simultaneously both too instinctual and too befuddled by maxims distilled from Rousseau by four or, arguably five, adolescents. Whether Joy is an actual person or a split mental projection generated by Michael's jealousy and his vexed relationship with Isobel, is an unresolved moot point. The painted Eyes, the written words and phrases with which the trees are adorned (tellingly 'FRATERNITY OR DEATH'; 'WHORE', or, 'down the slender trunk of a young birch: "STRENGTH THROUGH JOY"') (200), the rules and the ruthless verdicts on traitor-Citizens, the guillotine: this arboreal Republic becomes a grim parody of the worst excesses of the French Revolution, which has been an inspiration. However, what ensues is also a re-enactment of the ideological premises and cultural dichotomies implemented by the Enlightenment: the power of language (logos), the weight of surveillance (the Self and the Social Eye), the Order of Time (on the clock), Nature as Other, the split human psyche ('two selves dwelling in the same mind. The nightself and the dayself. The sleepself and the wakeself') (177). Michael's mental landscape is a schism that veers between two extremes. On the one hand, Joy (or, more credibly, Michael himself, projected as Joy) implements: Symbolic Order, Rules, Regulations, Sex, Naming, Shaming, Inquisitions, watching and indicting, control, punishment and death. On the other hand, when in the trees, Michael experiences a different, more pulsional, mode of being:

> Out beyond the realm of the eyes, I started to climb. I could feel the trees' energy pulsing in my fingertips as I clung to them. Their branches cradled me, gave me spring and leverage. It took me a little time to discover my rhythm [...] so my head was not filled with ideas and words [...] I climbed to the top of a beech [...] the horizon was clear and I could see the whole [...] the mountains looked flat [...] yet there was a living gleam to their brown flanks [...] They seemed to breathe and pulse [...] Ideas and words, republics and gods, all withered in their shadows [...] I thought of the ridiculous eyes I had painted, all those ridiculous words, and I laughed. The trees, the mountains and the buzzards laughed, too. (220)

Although it cannot be sustained in his situation, Michael's communion with the trees, mountains, birds and sky recuperates the animism in the natural world that has long been denied. 'Biosemiotics', that is, 'semiotic processes in nature' and 'ecosemiotics', 'semiotic processes between nature and human culture' (Maran, 2006, 457), are terms posited and discussed by Timo Maran in his essay 'Where do your borders lie? Reflections on the semiotical ethics of nature'. He acknowledges that 'most ethical theories, either anthropocentric or biocentric, fail to recognise the role of direct relations between a human subject and natural phenomena' (456). Maran's aim is 'to contribute to the rehabilitation of such relations, to show their fundamental role for our subjective realities and representational practices' (456). Humans are 'not only cultural but also biological creatures who can easily relate to semiotic activities on non-human living beings' (463). Taylor's description of Michael's experience up in the trees is an account of just such reciprocal relations.

Cosimo, living in eighteenth-century Italy, also takes to the trees in a spirit of adolescent rebellion, after defying his father, the Baron, at the dinner table. Cosimo, only newly admitted to the adults' dining room, rebels against his father's authority, and the accumulated injustices inflicted by him, by refusing to eat a dish of snails prepared by his sister. This rebellion entails climbing into a holm oak outside the dining room windows and vowing, in the face of all entreaties to rejoin the household, never to set foot on earth again. Cosimo does live a rich, mobile and long life while entirely living 'on the trees' (213).

The region in which the family home, Ombrosa, is located is so thick with trees, that Cosimo is able to roam without bounds or fear of trespass. As he explains to Viola, the daughter of the neighbouring Ondivaras, a family of nobles who compete with the di Rondos for feudal rights and pretentious distinctions, only the ground constitutes property: 'it's the ground that's yours, and if I put a foot on it, that would be trespassing. But up here I can go wherever I like' (19). Indeed, Cosimo travels far and wide, including to Olivabassa in Spain, where he joins a community of exiles, 'a whole tribe of Spaniards living on the trees' (125). This is where Cosimo conducts his first love affair, although later he is reunited with his grand passion, Viola. Cosimo's arboreal life consists of a whole spectrum of activity, including continuing his lessons, reading, hunting, sewing, building and engineering, distributing stolen goods to the poor and needy, holding salons and cultivating the minds of others including that of a notorious brigand, writing treatises and telling stories, keeping a companionate dog, bee-keeping and tree-pruning (for a small fee, but for the good of the trees, the fruitgrowers, those who had 'trees planted for shade or ornament' (103) – and himself):

> In fact, his love for this arboreal element made him […] help growth and give shape. Certainly he was always careful when pruning and lopping to serve not only the interests of the owner, but also his own, as a traveler […] And so, these trees of Ombrosa which he had already found so welcoming, now, with his newly acquired skill, he made more directly helpful, thus being at the same time a friend to his neighbour, to nature and to himself. (103)

Although too fanciful to be a programmatic guide to realise a utopia, there is something exemplary about the liberty, fraternity and self-determination that Cosimo achieves from his dwelling on the trees. However, from the outset, the novel is also a lament and an elegy for things that have been lost, not least, the treed landscape itself. The passage cited above, which shows the well-roundedness of Cosimo's citizenship, closes with a temporal reflection: 'Then, with the advent of more careless generations, of improvident greed, of people who loved nothing, not even themselves, all was to change, and no Cosimos will ever walk the trees again' (103).

Thus, writing in the 1950s, Calvino uses the eighteenth century as the locus for saying adieu, not just to a treed landscape but to the care and notice people took of it. In the early chapters of *The Baron in the Trees*, he provides some detailed context for the steady decline and disappearance of the panoply of indigenous tree species. Legend tells of being able to travel across countries, from Italy to Spain, tree to tree and monkey-like, without ever touching the earth. Of especial renown was the heavily treed gulf of Ombrosa, but due to lack of respect, greed and improvidence and to a utilitarian view of trees as timber for ship- and house-building, 'the world of sap' (29) was destroyed:

> Nowadays these parts are very different. It was after the arrival of the French that people began chopping down trees as if they were grass which is scythed every year and grows again. They have never grown again. At first we thought it was something to do with the war, with Napoleon, with the period. But the chopping went on. (28–29)

Calvino carefully notes the various species of trees and their attributes; it is clear that he is familiar with many of the trees he describes (Calvino's own parents were botanists, and he was an admirer and close observer of trees and plants.). He also notes the changes wrought in the eighteenth-century landscape by plant importers, botanists, taxonomists and landscape designers, as testament to the Enlightenment quest for knowledge and acquisition, and so integral to European domestic and colonial power. The garden at Ondarivas, for example, was full of 'the finest botanical rarities from the colonies' (15):

> For years boats had unloaded at the port of Ombrosa sacks of seeds, bundles of cuttings, potted shrubs and even entire trees with huge wrappings of sacking around the roots; until the garden […] had become a mixture of the forests of India and the Americas, and even of New Holland. (15)

It is not just in the Parks and gardens that indigenous trees and plants are replaced. The wider landscape has been affected too, 'since men have been swept by this frenzy for the ax' (217):

> And the species have changed too; no longer are there ilexes, elms, oaks; nowadays Africa, Australia, the Americas, the Indies, reach out roots and branches as far as here. What old trees exist are tucked away on the heights; olives on the hills, pines and chestnuts in the mountain woods; the coast down below is a red Australia of eucalyptus, of swollen India rubber trees, huge and isolated garden growths, and the whole of the rest is palms, with their scraggy tufts, inhospitable trees from the desert. (217)

At the end of the novel, it is clear that Ombrosa, the ancient and once heavily forested, 'a free commune for some time, tributary of the Republic of Genoa' (54) is simply 'no place' (from Gk *ou* not, *topos*, a place) (OED): 'Ombrosa no longer exists' (217) explains the narrator, Biagio. 'Looking at the empty sky [bereft of trees], I ask myself if it ever did really exist' (217).

The history of nature, trees, forests and forest management shows that the ideologies and epistemologies of the Enlightenment, and their legacies, are now ubiquitous in the West and much of the rest of the world. Across the globe, environmental crises mean that science (and commerce) are the touchstones for a sustainable global future, but environmental scientist, Daniel Botkin, says this approach simply will not prove effective; what is imperative is to find new metaphors to replace old ones in thinking about nature.

> We analyze ecological systems looking backward, as though they were nineteenth-century machines, full of gears and wheels, for which our managerial goal, like that of any traditional engineer, is [a] steady-state operation. To us, the mechanical view of Earth, nature as a machine seems an old and permanent one. But it is not. The mechanical Earth is a seventeenth-century idea that developed in the eighteenth-century, blossomed in the nine-teenth, and carried into the twentieth. (Botkin 2012, 16)

Botkin explains that one of the purposes of his book, *The Moon in the Nautilus Shell* (2012), is to facilitate a new 'perception' that we 'have not yet made our own' (4): the 'tight integration of people and nature' (4). 'This transition – from seeing ourselves as outside of nature and therefore only harming it, to seeing ourselves as within and part of nature' (4), involves the dismantling of 'the old machine idea that we have become accustomed to using for the past 200 years' (327).

It is not surprising that novelists depicting revolution and the desire for a new order should continue to engage with Rousseau as one of Europe's most influential thinkers, but what is also of note is the centrality of trees, both as a material reality *and* as utopian myth and metaphor in the two novels considered here. Rousseau, of course, loved a solitary walk in the woods, and in his meditations on how to recover 'natural man', the forest setting was often a trope. As Robert Pogue Harrison notes, Rousseau's 'state of nature' has 'the primeval forest' as its wholly imaginative or 'truthfully intuited' setting: 'Intuition enables him to imagine "natural man" wandering solitary through the great primeval forests of the earth, living a simple, innocent, and most importantly, *happy* life' (Harrison 1992, 128). Harrison also cites a lesser known treatise, Rousseau's *Projet de Constitution pour la Corse* (1765). Corsica's abundant woodland will provide amply for 'construction and heating' and, cautioning against the improvident tree-felling that has taken place in Switzerland and France, for instance, he recommends establishing 'early on an exact policemanship of the forests and to regulate cutting in such a way that reforestation equals consumption' (Rousseau in Harrison 1992, 126). The voice of 'the poet of nostalgia' sits irreconcilably alongside the same voice's 'reduction of nature to mere resource for enlightened exploitation' (Harrison 1992, 127). Harrison's term for Rousseau's double-ness is 'dividual': 'He is fraught with contradictions not only for those who reckon with the coherence of his doctrines, but above all, within himself, especially when it comes to his discourse about forests' (125). Rousseau's contemporary, the Encyclopaedist and warden of the Park of Versailles, Monsieur Le Roy was less ambivalent, it seems, and demonstrated the profound shift taking place:

> In Le Roy's article forests are stripped of the symbolic density they may once have possessed. They are reduced to the most literal of determinations, namely 'a great expanse of woodlands … composed of trees of all sorts.' Le Roy never once mentions the issue of wildlife. *The forest as habitat has disappeared*. If habitat is not an issue for Le Roy it is because the forest has already been conceived of in terms of timber. This timber, in turn, has already been conceived of in terms of its use-value. Use-value, in turn, has been linked to the concept of 'rights' – the rights of the state, the rights of the private owners, and the rights of posterity. (Harrison 1992, 121)

What were trees, or even individual species, are now 'timber', what was 'natural' is now 'managed'; what was 'common' (at least in terms of access and gleaning) is now 'property'. Forest historian, Nancy Langston notes, first, that 'it is a reasonable general-ization to state' that hitherto forests had been largely communal:

> No single person owned the rights to a forest; unlike agricultural lands, which did tend to be individual private property in many cultures, forests had broader – but not unrestricted – access. Customary tenure systems traditionally regulated access to common property resources within a forest such as fuel wood, grazing and what foresters now awkwardly term 'non-timber forest resources' such as berries and game. (23)

Second, and just as significantly, she explains that the Scientific Revolution and Enlightenment, 'tried to reduce the messy, fertile complexity of myth and undergrowth (Langston 2005, 24), and, by this, forests have become the repressed 'shadow of civiliza-tion' (Harrison). 'Forests unsettle, they overturn stability, they confuse clear distinctions [...] They are places of transformation, places where the human and the wild meet and get entangled in a web of myth, ritual, stories, worship and fear' (Langston 2005, 24). Thus, the forest is the compelling locus for, simultaneously, utopian *and* dystopian human-imagining. For Botkin, however, messy complexity *and* modern technology, 'an amalga-mation of the organic metaphor with a new technological metaphor' (327), is precisely what will generate new metaphors and the 'new understanding of the biological world' (327) that both Calvino and Taylor show us is needed.

'Nature deficit disorder'[1]

The politics of utopias, and their often polemical representation, have been much debated, not least for the way in which representations abound more than actual utopias. Discussions concerning the political and social *efficacy* of utopian fantasy ('a dream incapable of attainment') (Moylan 1986, 9) versus dystopian cautions ('narrative that images a society worse than the existing one' (Moylan 1986, 9) are also of long-standing. The two texts under consideration here, the one typified by dream and the other by nightmare, do not, as has already been noted, introduce the reader to well-delineated alternative worlds, whether fruitful or failed. Rather, they use the utopian tradition, particularly in relation to new Republics, and new relations, as a point of reference. In this regard, they are well aligned with Ruth Levitas's explanation of a 'utopian method' that is 'primarily hermeneutic':

> Read in this way, utopia does not require the imaginative construction of whole other worlds. It occurs as an embedded element in a wide range of human practice and culture. [...] We can explore culture (in its broadest sense) for its utopian aspects, its expression of longing and fulfilment [...] it is evident that contemporary culture is saturated with utopianism, even (or especially) where there is no figurative representation of an alternative world. (Levitas 2013, 5)

These texts might also fulfil the definition of 'the critical utopia' (Moylan, 10), which Tom Moylan cites as emerging after the 1960s and typical in the 1970s. Calvino's text of the late 1950s and Taylor's published in 2005, both referring to the utopian and republican idealism of the eighteenth century, can be seen as critical utopias where 'a central concern is an awareness of the limitations of the utopian tradition, so that these texts reject utopia as a blueprint while preserving it as dream' (Moylan 1986, 10). Ruth Levitas also notes that 'the "critical utopia" does not offer an unequivocal alternative' (111). For Moylan, she says, 'the critical utopia' is 'critical in three senses: it implies an Enlightenment sense of critique; a postmodern self-reflexivity and provisionality; and a political critical mass leading to change – the latter existing by implication outside the text' (111). For herself, more pessimistically, it matters more that 'the political impetus and intent of the critical utopia is not necessarily matched by political effectiveness' (111):

> A politically quiescent context and reading enables them to function only as critique. On the other hand, the very reflexivity and provisionality means that they can be seen as examples of utopia as method – the self-conscious promotion of interrogation of possible alternative futures from a position which registers both the necessary indeterminacy of the future and the plurality of the agents who will create it. (111)

I would argue, however, that these two texts can and should be read in a current context that is anything but 'politically quiescent' and that the 'indeterminacy of the future' is part of an urgency that increases and, perhaps uniquely, an indication of the plurality of agents that will be needed. As 'critical utopias' Calvino and Taylor argue, not just for a nostalgic past of a forested landscape, or for a recuperation of pre-Enlightenment perspectives on what we now call the 'biosphere' or 'environment', but new metaphors and ultimately, new ways of being in the biological and technological world. Although Calvino, certainly, and to a certain extent, Taylor, cannot have anticipated the context within which these books can now be read, they have both discerned and dramatised the ideological "developments" with which the world now grapples and seeks to ameliorate.

Calvino noted in 1980 that, despite his sense of particular place, he had 'always tried to avoid' 'a regressive local feeling' (viii). Taylor, too, speaks of and resists the recent trope of the Provencal idyll, along the lines first established by Peter Mayle in *A Year in Provence* in 1989. Neither of these texts are explicitly idealising a pre-industrial world, but they are both acknowledging profound damage and a severe 'deficit', instituted to a large extent and somewhat ironically, by the utopian ideals of the Enlightenment. Cosimo's rebellions enact an Enlightenment refusal to acknowledge inherited privilege, ancestral pretentions and arbitrary power. For Calvino, writing in the mid-twentieth-century context of the Cold War, of Soviet tanks entering Budapest, and 'the disintegration of the left-wing intelligentsia as a politically active collective subject in the 1950s' (Bolongaro 2003, 129*), The Baron in the Trees* is a meditation on 'the problem of the intellectual's political commitment at a time of shattered illusions' (Calvino 1980, x). Eugenio Bolongaro extends the comparison: 'Biagio's farewell to the hopes of the Enlightenment echoes Calvino's own farewell to the hopes for social change that the resistance had raised' (Bolongaro 2003, 128).

Instead of polemic, or material realism of his first novel, Calvino's 'republican' novel is a fable. Of its meaning, the former resistance fighter and Communist Party-member was vague and non-prescriptive: 'the reader must interpret the stories as he will, or not interpret them at all and read them simply for enjoyment' (Calvino 1998, ix). Genre is always a part of a wider political perspective, and it is an aspect about which Calvino wrote a great deal, at first rather defensively, and then with a greater confidence in the appropriateness of allegory, fable and complexity. As Calvino explains in the 'Introduction' to *Our Ancestors*, which brought together in one volume three 'romances' of which 'The Baron in the Trees' was one, and which had 'raised eyebrows' when first published in the 1950s:

> In the fifties, [...] what was expected from Italian literature in general, and from me in particular, was novels [...] From my first published works I was thought a 'realistic' writer; indeed, a 'neo-realistic' one [...] I was not prepared for the outcry that greeted it or the fact that it would be thought a shocking turnabout on my part. The fact is that after my first novel,

written in 1946 [...] I had made efforts to write the realistic-novel-reflecting-the-problems-of-Italian-society [...]. (At the time I was what was called a 'politically committed writer'). (vii)

Almost abandoning writing instead of trying to write what he *'ought* to write', he began 1951, 'doing what came most naturally to me – that is, following the memory of the things I had loved best since boyhood' (Calvino 1998 [1980], vii). Included in these were narratives written by exemplars such as R.L. Stevenson, J.L. Borges, Voltaire and the nineteenth-century Italian novelist, Ippolito Nievo (a particular influence for *The Baron in the Trees*). An 'experience' common to all the three tales ('The Cloven Viscount', 'The Baron in the Trees' and 'The Non-Existent Knight'), however, is Calvino's memory of 'landscape':

> My own experience appears in the three stories; to start with, in their landscape. Although they take place in imaginary countries [...] they breathe the air of the Mediterranean which I had breathed throughout my life. Much of Italian literature has regional roots [...] my own literary history starts from a very particular place. My home was in San Remo on the Riviera [...] which building development in the postwar years has now made unrecognizable [...] So the Ligurian landscape, *where trees have almost disappeared today, in the Baron is transformed into a kind of apotheosis of vegetation* [my emphasis]. (Calvino, *Our Ancestors,* ix)

While there has been a good deal of discussion about Calvino's personal (and textual) politics, less is known about Taylor's. This leads, perhaps, to a too ready evaluation of *The Republic of Trees* as simply a dystopian shocker. ('The book is a fable, a gothic fiction, and [...] like all good fairy tales, pretty disturbing') (Taylor, *The Observer*, 2005). In an article entitled 'Daydream Believer', in *The Observer* in 2005, Taylor recounts his own childhood experience and memories, which are at odds with those of Calvino, but which are also clearly important to the novel. Taylor's experience, translated into Michael's, was of the antithesis of the 'apotheosis of vegetation'; an 'apotheosis' of asphalt, traffic and paving stones, perhaps. He recounts his own experience as a child: 'I grew up in a suburban housing state in the Midlands and used to dream about running away to the forest' as well as an adult one, as the father of a young child:

> One spring morning five years ago, I walked our eldest son Oscar to school in the Hertfordshire commuter village where we lived. It was a nice place – there was a nature reserve and a canal – but as usual we found it impossible to have a conversation over the roar of the traffic. On the way back I saw a house [... in France] in an estate agent's window [...] I was The Observer's pop critic at the time [...] A dream of escape began to form in my mind. (Taylor 2005a)

Just as the elderly Biagio, at the end of *The Baron in the Trees*, laments the loss of a once-green landscape ('for us [...] used to living under those green domes, it hurts the eyes to look out now') (217), so too, does a young suburb-dwelling Michael Vignal, two centuries later, regret the loss of what must have been:

> Two hundred years earlier, it would have been part of a great forest, but now the only trees were pruned dwarfs, planted at regular intervals next to the road – little slave-trees, kept to remind all those managers and accountants and salespeople of the vast green wilderness crushed beneath their patios. (3–4)

Michael's rebellions are darker and more dreadful than Cosimo's aerial ones; Cosimo remains above the ground, poised between earth and sky, and at the end of his life, ascends into the air. Michael's deeply divided self is emblematised by the stark duality of his personality depending whether he is in the trees or on the ground. Cosimo's death entails being airborne after a long and busy life living in trees. Michael's, we believe, is about to take place by falling disastrously to the ground. These two endings could be seen to be emblematic parodies of the conventional extremes of the unrealistic idealism of utopias, on the one hand, and the anarchic chaos of dystopias on the other. Cosimo's triadic and holistic epitaph: 'Lived in trees – Always loved earth – Went into sky' (216) has its antithesis in Michael's murderously maniacal 'Kill him now' and his suicidal fall from 'the very edge of an oak branch': 'In the euphoria of the moment, I lost my balance' (223). The losses in Michael's short life are extreme (the death of his father by electrocution after severing a lawnmower cable in a suburban garden; the suicide of his mother, slitting her wrist when her now-fatherless children were at school; the abusive treatment by the French aunt); there is more to indict in his situation than the lack of accessible woodland. Nevertheless, the lack of accessible woodland has something to do with the wider social demarcations and deprivations typified by the lives of Michael and his family. There is no doubt that there is much grist here for the analysis of post-1980s Britain, and indeed, to some extent, France. Taylor, born in 1976 and growing up in the 1980s, his American Studies university degree in the early 1990s, a job as an on-the-road popular-music journalist, his unexpected redundancy, his move with his young family to France in the hopes of a better lifestyle – all inflect an understanding of what might be the contextual correlative to Calvino in the 1950s. Both authors indicate a problematic 'utopia': the isolate and independent self. Cosimo's insistence that 'there can be no love if one does not remain oneself with all one's strength' results in the breakdown of his relationship with Viola, who retorts 'Be yourself by yourself, then' (177). Similarly, Taylor's suggests a reason for Michael's murderous removal of rivals: 'Perhaps what he wants is the perfect society, which would be a society of one' (Taylor 2005b). However, what Cosimo and Michael as post-Enlightenment and post-industrial human subjects really lack, and only at times recuperate by actually (and impractically) being 'in trees' is the sense of the integrity of a relationship with Nature and, as human subjects, part of Nature. This is the family with whom Cosimo and Michael need to restore relations.

The rights of trees

It is, perhaps, impossible to conjecture what utopia *for* trees might be like; what can more confidently be stated is that, largely, certainly since the Enlightenment and where human beings are involved, forests and trees are at a far remove from utopia in terms of the well-being or the rights of trees themselves. Western society is well aware of its impact on forests at home and abroad; the loss of rainforest is widely lamented and albeit late-in-the-day measures are taken to redress the damage and ensure more sustainable futures for forests (and for timber). What is less often addressed is how the ways humans have thought about forests, and their relation to (or lack of relationship with) forests, has its origins in 'discourses on inequality' that become foundational from the Age of Enlightenment onwards. Those who have considered the 'development' of forests and forest management do seem to be very clear where the problems originated, as for instance, James Scott, in *Seeing Like a State: How Certain Schemes to Improve the Human Condition have Failed* (1999):

Scott argues that from the rise of the modern state in the eighteenth century, those in authority have tried to organize society and ecosystems through centralized, topdown plans that simplify human and ecological connections, to further the state functions of taxation, conscription and maximization of the state's resources. Scott shows how centralized planning and high modernism have led to ecological and human disaster. (Langston 2005, 22)

Where Western society, in particular, may think it is taking good care of trees, there is still the prevailing sense that trees are objects to be managed. Poets Paul Farley and Michael Symmons Roberts discuss 'our English woods', the yearning for 'the original tracts of greenwood', albeit, now, only 'a complicated and sustaining myth' (Farley and Roberts 2011, 162). Instead, we have the worthy and, no doubt, well-meaning 'man-made greening' of 'our postmodern woodlands':

> You can easily get lost in the woods of mission statements and manifestos where community forests are concerned: the twelve woodlands established so far aim to 'deliver a comprehensive package of urban, economic and social regeneration'[...] 'creating high quality environments for millions of people by revitalising derelict land, providing new opportunities for leisure, recreation and cultural activities, enhancing biodiversity, preparing for climate change and supporting education, healthy living and social and economic development'. (Farley and Roberts 2011, 163–164)

Whatever ideals engender these 'mission statements and manifestos', they are at a far remove from 'wilderness', and at an even further remove from 'rights for trees and other forest inhabitants'. The fictional eighteenth-century Cosimo wrote, as already noted, a *Constitutional Project for a Republican City with a Declaration of the Rights of Men, Women, Children, Domestic and Wild Animals, Including Birds, Fishes and Insects, and All Vegetations, whether Trees, Vegetables, or Grass* (205) and with reference to it, Jonathan Bate exhorts: 'Imagine the species and the forests that would still be living if some of the principles of this project had found their way into the enlightened declarations of the American and French revolutionaries' (Bate 2000,169), and he goes on to cite Harrison: 'Today we are witnessing the consequences of those one-sided declarations of the right of a single species to disregard the natural rights of every other species' (169). That Cosimo's treatise remained a 'dead letter', even in the context of various declarations and vindications of rights is, states Bate,

> because the Enlightenment style of discourse – the systematic, theoretical 'project' – encouraged such hierarchical descriptions of nature and did not afford 'a way of thinking which allowed for the possibility of extending the universal rights of man into a proclamation of the universal rights of nature.' (170)

At the Ecuadorian Constituent Assembly of Montecristi, held in 2008, a Declaration of the Rights of Nature *was* announced with the aim of incorporating it into its national Constitution. For Alberto Acosta, Chair of the Assembly at the time, minister of energy and a presidential candidate in 2013, this acknowledgement constitutes a radical repositioning of post-Enlightenment perspectives: 'that is, by perceiving Nature as a subject of rights, and, moreover, investing it with the right to be restored when destroyed, a milestone was established for mankind' (Acosta 2010, 5). Acosta's article is extraordinarily apposite to the premises and contributions outlined above; a further surprise is that Calvino's Baron Cosimo di Rondò, and his treatise is invoked within it as a precursor to the 2008 Ecuadorian declaration: 'the idea of investing nature with rights even has its precedents in the western world. This thesis is already reflected by Italo Calvino' (Acosta

2010). Jonathan Bate finds that Cosimo (and Calvino) fail to launch and implement the declaration because:

> the universal rights of nature cannot effectively be *declared* in a systematic treatise; they can only be *expressed* by means of a celebratory narrative. They require not an Enlightenment project but a Romantic riot of sketches, fragments and tales – narratives of community, reminiscences of walking and working, vignettes of birds and their nests, animations of children and insects and grass. (Bate 2000, 170)

What the Ecuadorian Assembly can now dare to posit in these terms, could only be gestured at by a twentieth-century fiction. Although some of the responses to the Ecuadorian Declaration were damning ('nonsense' and 'conceptual gibberish'), Acosta retorts that 'throughout legal history, every broadening of rights was previously inconceivable' (Acosta 2010). The move from anthropocentrism to a biocentrism specifically addresses and aims to deconstruct those ideologies privileged in Europe in the eighteenth century: that mankind dominates Nature and sees it primarily as a material resource for economic growth. Nature, says Acosta, 'a term conceptualised by humans, must be wholly reinterpreted and revised if we intend to preserve the life of human beings on the planet' (Acosta 2010, 1).

> The task is to investigate and talk with Nature, always appreciating that we are immerse [sic] in it. Thus, consolidation of a new form of interrelation of human beings with Nature is required, as being an integral part of it. This entails having a scientific understanding of Nature, and at the same time, admiration and reverence for it, an attitude of identification with Nature, far from ownership and dominance, and very close to curiosity and love. (1)

'Curiosity and love', interrelationship and intimacy is certainly depicted in passages in both novels, where the protagonists benefit from a dynamic reciprocity with trees and nature more broadly, although Cosimo has the edge in this regard:

> [...] he who spent his nights listening to the sap running through its cells; the circles marking the years inside the trunk; the patches of mold growing ever larger [...] the birds sleeping and quivering in their nests [...] and the caterpillar walking and the chrysalis opening. There is the moment when the silence of the countryside gathers in the ear and breaks into a myriad of sounds: a croaking and a squeaking, a swift rustle in the grass, a plop in the water, a patterning on the earth and pebbles [...] The sounds follow one another, and the ear eventually discerns more and more [...]. (70)

Not only does Cosimo have an affinity and respect for nature, he has, by virtue of his unique situation, the opportunity to learn more about a world from which his own species is increasingly and irrevocably, it seems, estranged. For Michael, a more-desolate child of the late twentieth century, the estrangement and general sense of collapse is too severe for his communion with nature to help him.

The world of sap

Utopian visions and political idealism are often seen as unrealistic and, while instructive and in their way, at a remove from the pragmatism that real life commands. In terms of clear political agendas and well-described alternative new orders, both *The Baron in the Trees* and *The Republic of Trees* fail; in terms of literary utopias they are hybrid, open-ended and parodic. The 'neo-realist' Calvino wrote of his three stories; the tales are 'born' 'from an image, not from any thesis I want to demonstrate' (ix) and that 'around the image extends a network of meanings that are always a little uncertain, without insisting on an unequivocal, compulsory interpretation' (ix). One could say that, like the forest, these two forest-located and republic-referenced novels unravel the compulsory and systemic binaries that eighteenth-century installed: 'it is these distinctions and dualisms that the wild forest continually confuses' (Langston 2005, 24). To read the Ecuadorian Declaration of the Rights of Nature is to wonder at the sentiment and rhetoric but to remain sceptical, perhaps, about how feasible and far-reaching such principles can be. However, Acosta is not simply a utopian visionary; although considered 'crazy by some of his cabinet peers' (Vidal, *The Guardian,* 2010), Acosta, in 2010, recommended leaving oil in the ground, as opposed to drilling it and earning his country in the region of $7 billion: 'against that must be put the incalculable cost of climate pollution, of trashing the Amazon rainforest', 'the conflict and devastation it will cause in one of the most diverse regions of the world' and the inevitable extinction of 'two uncontacted tribes believed to be in the vicinity' (Vidal). An acknowledged inspiration for Acosta, Cosimo's once 'dead letter' may actually be coming to life, and 'the world of sap' and all its inhabitants might mean more than a now obviously obsolete and destructive model.

Note

1. A term coined by Richard Louv and discussed in *Last Child in the Woods* (2005).

References

Acosta, A. 2010. "Toward the Universal Declaration of Rights of Nature: Thoughts for Action." *For AFESE Journal*, August 24. Accessed November 8, 2013. http://therightsofnature.org/wp-content/uploads/pdfs/Toward-the-Universal-Declaration-of-Rights-of-Nature-Alberto-Acosta.pdf

Bate, J. 2000. *The Song of the Earth*. London: Picador.

Bennett, J. 2010. *Vibrant Matter: A Political Ecology of Things*. Durham, NC: Duke University Press.

Bolongaro, E. 2003. *Italo Calvino and the Compass of Literature*. Toronto, ON: University of Toronto Press.

Botkin, D. B. 2012. *The Moon in the Nautilus Shell: Discordant Harmonies Reconsidered*. Oxford: Oxford University Press.

Calvino, I. 1959 [1957]. *The Baron in the Trees*. Translated by Archibald Colquhoun. London: Harcourt.

Calvino, I. 1998 [1980]. *Our Ancestors*. London: Vintage.

Calvino, I. 2013. *Italo Calvino: Letters, 1941–1985. Selected and Intro. Michael Wood*. Translated by Martin McLaughlin. Princeton, NJ: Princeton University Press.

Cornwell, J. 2005. "Seeing the Trees." *The Australian*, 28 May. Accessed January 18, 2013. http://www.sam-taylor.com/republic/cornwell.htm

Farley, P., and M. Symmons Roberts. 2011. *Edgelands: Journeys into England's True Wilderness*. London: Jonathan Cape.

Harrison, R. P. 1992. *Forests: The Shadow of Civilization*. Chicago, IL: University of Chicago Press. http://www.samtaylor.com/books.htm#republic_reviews (accessed September, 2013 – website since closed).

Langston, N. 2005. "On Teaching World Forest History." *Environmental History* 10 (1): 20–29. doi:10.1093/envhis/10.1.20.

Latour, B. 2004. *Politics of Nature: How to Bring the Sciences into Democracy*. Cambridge, MA: Harvard University Press.

Levitas, R. 2013. *Utopia as Method: The Imaginary Reconstruction of Society*. Basingstoke: Palgrave Macmillan.

Louv, R. 2005. *Last Child in the Woods: Saving our Children from Nature-deficit Disorder*. London: Atlantic Books.

Maran, T. 2006. "Where Do Your Borders Lie? Reflections on the Semiotical Ethics of Nature." In *Nature in Literary and Cultural Studies: Transatlantic Conversations on Ecocriticism*, edited by C. Gersdorf, and S. Mayer. Amsterdam: Rodopi.

Monbiot, G. 2013. *Feral: Searching for Enchantment on the Frontiers of Rewilding*. London: Allen Lane.

Moylan, T. 1986. *Demand the Impossible: Science Fiction and the Utopian Imagination*. London: Methuen.

Sargisson, L. 2013. *Fool's Gold: Utopianism in the Twenty-First Century*. Basingstoke: Palgrave Macmillan.

Scott, J. C. 1999. *Seeing Like a State: How Certain Schemes to Improve the Human Condition Have Failed*. New Haven, CT: Yale University Press.

Taylor, S. 2005a. "Daydream Believer." *The Observer*, March 13. Accessed January 18, 2013. http://www.sam-taylor.com/republic/daydream.htm

Taylor, S. 2005b. "Does Joy Really Exist? And Other Questions". Accessed January 18, 2013. http://www.sam-taylor.com/republic/joyQ&A.htm

Taylor, S. 2005c. *The Republic of Trees*. London: Faber & Faber.

Vidal, J. 2010. "Oil: Can Ecuador See Past the Black Stuff? A Revolutionary Plan to Leave Ecuador's Abundant Oil in the Ground Could Show the World Just What's Possible." *The Guardian*, September 28. Accessed March 1, 2013. http://www.guardian.co.uk/global-development/poverty-matters/2010/sep/28/ecuador-oil-extraction-amazon-yasuni

Hope of a hopeless world: eco-teleology in Margaret Atwood's *Oryx and Crake* and *The Year of the Flood*

Nazry Bahrawi[a,b]

[a]*Humanities, Arts and Social Sciences, Singapore University of Technology and Design, Singapore;* [b]*Middle East Institute-National University of Singapore, Singapore*

At the hands of Margaret Atwood, literary ecological tropes assume a dystopian demeanour. Through a comparative analysis of her related speculative fictions, *Oryx and Crake* and *The Year of the Flood*, this paper argues that Atwood's ecocriticism is a desecularised manifesto that imagines a messianic form of ecotheology. It does so by first outlining expressions of 'overhumanisation' that act as Atwood's critique of scientism. This essay makes the case that Atwood's ecotheology is better figured as *eco-telology* that works by apophasis so as to articulate hope in a hopeless world. It concludes that Atwood's eco-teleology postulates the human subject as a 'thing in between' as theorised by David E. Klemm and William Schweiker, and gestures to Ernst Bloch's idea of the 'Not-Yet' to feed its utopian desire.

Introduction

The human-sparse, science-torn world depicted by the Canadian author Margaret Atwood in *Oryx and Crake* and *The Year of the Flood* strikes at the heart of the twenty-first century public debate on climate change as an anthropogenic phenomenon. Published in 2003 and 2009 respectively, the novels are premised on the idea that unbridled human agency has indeed led to ecological abuse, even though Atwood has not explicitly articulated such an opinion in public.

While a sociological enquiry can better address the politics behind climate change policy, this essay explores the bioethical facets of Atwood's novels stemming from the speculation that human flourishing in the form of scientism has ironically led to the manipulation and degradation of nature. Yet it must first be qualified that the novels are far from univocal. By way of literary texture, *Oryx and Crake* is imbued with an over-arching sense of wryness that can be gleaned from the comical yet uncanny feel of the repertoire of its 'new-old' bio-beings like the ChickieNobs, which are chicken designed with a 'sea-anemone body plan' but look like 'a large wart' (Atwood 2004, 202–203). If *Oryx and Crake* takes its readers on a carnivalesque tour of the MaddAddam realm, *The Year of the Flood* navigates them through its violent underworld, rendering it much darker in tenor.

Despite this distinction, the apocalyptic world of these first two novels of the MaddAddam trilogy does collectively imagine science as having enabled the creation of an incurable virus that had wiped out most of humanity, and in the process littered the world with dangerous hybrid creatures such as the liobam, a genetically engineered creature that has the harmless appearance of a lamb but the hunting ability and razor-sharp incisors of a lion. Nature, used and abused, has returned to wreak havoc on earth with a vengeance.

Such perversions conform to what theologians David E. Klemm and William Schweiker describe as 'overhumanisation' in *Religion and the Human Future: An Essay on Theological Humanism* (2008). Premised on the belief that humanity is 'the master of its own destiny and source of its own moral law' (Klemm and Schweiker 2008, 157), 'overhumanisation' can be seen as the dark side of secular humanism, a philosophical outlook that has enabled progress but also resulted in some of our most daunting problems. Klemm and Schweiker write of its excesses:

> This project [secular humanism] has brought advances in knowledge, the lessening of diseases and want, and the formation of freer and more open democratic societies. Yet it had also led to the profaning of life through wars, ecological endangerment and cultural banality. Part of overhumanisation is also the unjust distribution of its goods – say, medicine, clean water, stable social orders – and the unfair distribution of destructive forces of modern societies: pollution, environmental damage, lack of access to hi-tech resources, astonishing poverty. (2008, 14)

To be sure, 'unjust distribution' of resources is not a problem unique to modern times. It is, in fact, a defining feature of feudal societies that have not embraced secular humanism. Yet Klemm and Schweiker's point is not that this disparity is novel but that secular humanism as a benign project has not quite delivered on its promise. Unjust distribution of resources remains an issue, one that is possibly even enabled by secular humanism.

In Atwood's bifurcated alternate reality, the tension between human flourishing and human over-reach is symbolically imagined through the vastly disparate built environments of the Compounds and the pleeblands. The first refers to privileged societies where genius scientists create new bio-products and business people strategise ways to market them under the watchful eye of CorpSeCorps, a private security firm that manages these gated areas in an authoritarian manner. In contrast, life in the latter is less glamorous and more perilous. Simply put, the pleeblands are ghettoes rife with diseases, criminal activities and poverty.

Yet the sense of irony that permeates the two novels also means that the Compounds and the pleeblands are not easily categorised into 'desirable' and 'undesirable' habitats. The Compounds may be the epitome of good living but they are also where the 'BlyssPluss' supervirus first saw the light of day. That is to say, the 'good place' is where a tool of genocide was forged. Meanwhile, the hazardous pleeblands are also where art truly thrives. The visual artist Amanda Payne uses the pleeblands to practice her innovative form of art called vulturising, in which she photographs vultures feeding on dead animal parts arranged into words. On the one hand, the case of Amanda can be read as a satirical take on modern art as hyperbolic and pretentious. On the other, it can be interpreted as complicating the idea that the pleeblands are undesirable. To push this later idea further, the pleeblands are also where members of the environmentalist sect who call themselves the God's Gardeners survive the hazards of the post-apocalyptic MaddAddam world, most clearly expounded through the perspectives of Ren and Toby in *The Year of the Flood*.

This essay also argues that the novels have most profoundly recouped the validity of 'faith' in an empirical world, specifically through their endorsement of a messianic form of ecotheology, or 'eco-teleology' as I term it. By 'faith', I do not simply mean organised religion, but the impulse to place trust in a world view or belief system that is non-empirical, and thus unverifiable. Seen this way, 'faith' includes not just the likes of Christianity but also Marxism, among others. Dystopic as they are, the texts posit that seco-teleology can save humanity from certain biological doom, though such a postulation is speculative more than prophetic. Writ large, this can be taken to mean that 'faith' in the MaddAddam realm is its most pervasive utopian impulse, the *hope* of its hopeless world.

Even so, it must also be noted that the all-permeating irony of the two texts problematises the very idea of 'faith' itself. The liobam, for instance, may have been made possible by the scientific method of genetic modification but it was really the brainchild of a group of religious fanatics known as the Lion Isaiahists, whose members believe that this hybrid creature was 'the only way to fulfill the lion/lamb friendship prophecy without the first eating the second' (Atwood 2010, 94). It is also crucial to acknowledge that the God's Gardeners as the other, more central, religious group in *The Year of the Flood*, is a messianic sect whose theism is built along Darwinian principles. For instance, its members are averse to CorpSeCorp's bioengineering or selective breeding in Darwinian terms, preferring instead to preserve the existing natural order encapsulated by their veneration of all things flora and fauna. This can be gleaned from the following excerpt in which a senior female Gardener Pilar tells Toby, one of the two protagonists from *The Year of the Flood*, about the importance of bees to the Gardeners:

> Pilar had a fund of bee lore. A bee in the house means a visit from a stranger, and if you kill the bee, the visit will not be a good one. If the beekeeper dies, the bees must be told, or they will swarm and fly away. Honey helps an open wound. A swarm of bees in May, worth a cool day. A swarm of bees in June, worth a new moon. A swarm of bees in July, not worth a squashed fly. All the bees of a hive are one bee: that's why they'll die for the hive. 'Like the Gardeners,' Pilar said. Toby couldn't tell whether or not she was joking. (Atwood 2010, 99)

Here, the text argues against the anthropocentric idea that nature's purpose is to serve humanity, an idea which environmental philosopher Patrick Curry in his book *Ecological Ethics* traces back to seventeenth-century Renaissance thinkers such as Francis Bacon and Robert Boyle (2011, 37). By postulating that humanity is not the superior race ('Like the Gardeners') but one among many earthly species of equal standing, the excerpt references the philosophy of Deep Ecology, which specifies – among its eight basic principles formulated by philosophers Arne Naess and George Sessions in 1989 – that 'humans have no right to reduce this richness and diversity [of life-forms] except to satisfy vital human needs' (Curry 2011, 101).

Most interestingly, the Gardeners' blend of Christian-Biology faith could itself be seen as the result of Darwinian natural selection in which aspects of Christian doctrines that help the Gardeners survive their harsh environmental conditions are *naturally* retained (like sainthood to mark eco-heroes worthy of reverence, among others) while non-relevant ones are discarded (for instance, the anthropomorphism of Jesus Christ).

Through the examples of the liobam and the Gardeners, the texts do not simplistically equate 'faith' to a sense of Godliness. Like the 'new-old' creatures of the MaddAddam realm, 'faith' too has been spliced into something between science and religion. Such a reading is congruent to the challenge that *Oryx and Crake* and *The Year of the Flood* pose to another kind of fixity – that is, between the literary genres of utopian and dystopian fictions. As Gerry Canavan argues, both novels conform to Atwood's notion of *ustopia* as 'a combination utopia-dystopia, where the two exist in simultaneous superposition with one another, each containing the other' (2012; Atwood 2011). Canavan may be talking about Atwood's works, but a similar albeit more general observation has been made earlier by Rafaella Baccolini when she writes: 'Utopia is maintained in dystopia, traditionally a bleak, depressing genre with no space for hope in the story; only *outside* the story: only by considering dystopia as a warning can we as readers escape such a dark future' (2004, 520).

Desecularising utopia

As works of speculative fiction, *Oryx and Crake* and *The Year of the Flood*, take a cynical view of scientism, defined by philosopher Tom Sorell as 'the belief that science, especially natural science, is much the most valuable part of human learning – much the most valuable part because it is much the most authoritative, or serious, or beneficial' (1991, 1).

In *Oryx and Crake*, the tension resulting from the split personality of Jimmy/ Snowman is reminiscent of the Roman god Janus whose two faces are simultaneously gazing at the past and the future. Indeed, Jimmy as Janus past inhabits the text in the form of a guide to the pre-apocalyptic MaddAddam world, a milieu wherein scientists were considered society's elites while artists led a proletarian existence. His counterpart Snowman, meanwhile, is introduced as Janus future, the sole human survivor of the post-apocalyptic MaddAddam world ravaged by the supervirus that has killed most of humanity, a world now occupied by the naïve bioengineered pseudo-human race known as the Crakers. Though the text's ending reveals other human survivors, Snowman's memories of his childhood for the most part of *Oryx and Crake* effectively outline 'overhumanisation' as a critique of scientism. The text describes Snowman's defamiliarisation with everyday objects and circumstances that once made sense to Jimmy – so much so that Snowman begins to hate 'these replays' (Atwood 2004, 68). The following excerpt, which describes Snowman's thoughts as he encounters a 'Men at Work' sign, is a case in point:

> Men at Work, that used to mean. Strange to think of the endless labour, the digging, the hammering, the carving, the lifting, the drilling, day by day, year by year, century by century; and now the endless crumbling that must be going on everywhere. Sandcastles in the wind. (Atwood 2004, 45)

Irony is once again detectable in the above passage if one considers its temporal aspect. As his future Snowman self, Jimmy is channelling the more ancient literary figure, Ozymandias, from Percy Bysshe Shelley's sonnet of the same name. On the statue of this 'king of kings' is inscribed the words 'Look on my works, ye Mighty, and despair!', serving a reminder to all who encounter it of the perishability of worldly things. In the same way, the above passage makes us think about the fleeting nature of the material world, though I argue that it takes this trope one step further. Here, the 'Men at Work' sign can be read as an allegory for progress on the back of wondrous scientific breakthroughs that punctuate Jimmy's world. As Janus future, Snowman has come to see the seedier side of progress as he mulls over the significance of the sign in positing the idea that science's 'endless labour' is no longer about bettering the here and now, as much as its unravelling, its deconstruction. One such breakthrough springs to mind – the pigoons. Before the outbreak, these oversized man-made pig hosts were used to harvest human organs 'that would transplant smoothly and avoid rejection' (Atwood 2004, 22). After the outbreak, they were Snowman's nemesis, always eyeing him as food, as 'a delicious meat pie just waiting to be opened up' (Atwood 2004, 268). Read this way, the defamiliarisation of everyday objects is also the defamiliarisation of science. If the labour of science was once seen as crucial to human flourishing, all that hammering, carving, lifting and drilling that have resulted in the pigoons and other bioengineered species are in fact symptomatic of overhumanisation. With the pigoons, the term 'overhumanisation' becomes ironically apt in the way these creatures resemble humanity in their behaviour. The following excerpt from *Oryx and Crake* in which Jimmy was being hunted down by the pigoons outlines the latter's ability to be conniving:

> They have something in mind, all right. He turns, heads back towards the gatehouse, quickens his pace. They're far enough away so he can run if he has to. He looks over his shoulder: they're trotting now. He speeds up, breaks into a jog. Then he spots another group through the gateway up ahead, eight or nine of them, coming towards him across No Man's Land. They're almost at the main gate, cutting him off in that direction. It's as if they've had it planned, between the two groups; as if they've known for some time that he was in the gatehouse and have been waiting for him to come out, far enough out so they can surround him. (Atwood 2004, 267)

In shedding light on the less flattering aspect of human progress, the MaddAddam novels speak to an existing train of thought that puts science on a pedestal, or the ideology known as scientism. Curry defines scientism as 'the modern cult of science, according to which science is not one way of being among many but the only valid or *true* one' (2011, 25).

When pitted against such tradition of thought, the decoupling of science from the ideology of progress that permeates the MaddAddam texts desecularises utopia. The alternate, better world will not culminate in a homogeneous civilisation defined by widespread empirical thinking, technological inventions and a culture of experimentations. Rather, salvation is invested in doctrines that could *not* be proven true such as the Darwin-centric theology of Adam One, as I explore below. The end of scientism is nigh.

Here, one can find some parallels between the novels' rebuke of human betterment through science and Curry's critique of scientism. Curry begins by questioning the entrenched idea that science can help humanity reach 'objective' truth about the world. He argues that the practice of science shows the opposite, writing that 'in a society dominated by financial, commercial and fiscal imperatives, science is no more immune than any other human enterprise to the corruption entailed by selling your services to the highest bidder' (Curry 2011, 26). To exemplify his point, Curry cites the real-life example of medical research in which pharmaceutical companies determine what is to be studied. Seen this way, Curry's critique of scientism, just as Atwood's MaddAddam novels, views empirical thinking suspiciously.

More importantly, Curry postulates that scientism threatens the possibility of ecological ethics in the way modern science sees nature as a 'material' and not 'spiritual' phenomenon. Curry believes that for humans to appreciate nature, we will first need to inculcate a sense of spiritual bond to the natural world. This requires cultivating a desecularised view of nature that upholds it as '*both* 'spiritual' object *and* 'natural' object' (Curry 2011, 138). Curry makes the important qualification that the desecularisation process is not anti-science but anti-scientism (2011, 25).

A similar mode of thinking can be observed in Atwood's novels. In *The Year of the Flood*, desecularisation is expressed through the God's Gardeners. Its members uphold the Green Bible as scripture, and make saints out of 'secular' personages like Terry Fox, the disabled athlete, and Dian Fossey, the murdered anti-poaching zoologist. This sacralisa-tion of the secular is most creatively presented in Atwood's repackaging of the genesis of Earth's creation, which hybridises the scientistic 'Big Bang' theory with the Christian account of God's intervention:

> Remember the first sentences of those Human Words of God: the Earth is without form, and void, and the God speaks Light into being. This is the moment that Science terms 'The Big Bang', as if it were a sex orgy. Yet both accounts concur in their essence: Darkness; then, in an instant Light. But surely the Creation is ongoing, for are not new stars being formed at every moment? (2010, 11–12)

In the MaddAddam reality, it is not just humans who are driven by the belief that a better world requires a sense of faith but so too the blue-skinned Crakers, programmed with the diminished ability to think conceptually on their creator Crake's reasoning that '[s]ymbolic thinking of any kind would signal downfall. … Next they'd be inventing idols, and funerals, and grave goods, and the afterlife, and sin, and Linear B, and kings, and then slavery and war' (Atwood 2010, 361). The Crakers' curiosity led them to develop a religion centred on Snowman's narrative that Crake is their creator, their God, while Snowman is their prophet. This may be an ironic comment on Atwood's part on the idea that creatures with the ability to think symbolically are not able to rid themselves of religious faith. With the Crakers, the text goes beyond mere desecularisation. It intimates to a sense of the post-secular, not one defined by the idea that 'faith' has returned after a period of absence, but that which suggests it has never really disappeared in the first place. In other words, the case of the Crakers provocatively stipulates that a proclivity to believe in something bigger than ourselves, to trust in a grand narrative of how the world works, to have 'faith', is almost the default position of humankind – even if humanity in the form of the Crakers is a filtered version of our current selves.

From ecotheology to eco-teleology

While cast within a dystopic mould, the first two MaddAddam novels encapsulate some semblance of hope, though this feature appears to be more prevalent in *The Year of the Flood* than *Oryx and Crake*, as I will soon illustrate. Before that, it must be qualified that this sense of measured hope does not make the novels literary utopias but renders them closer to 'critical dystopias'. Such a reading takes Baccolini's nuanced reading of dystopia as 'a warning' against the possibility of a certain dark future and Canavan's point about ustopia being a hybrid of utopia–dystopia a step further. Indeed, the MaddAddam novels are both these things. More precisely, both novels are dystopias 'informed and empowered by a utopian horizon that appears in the text – or at least shimmers beyond its pages' despite their 'pessimistic forays' – a definitive feature of 'critical dystopias' according to literary scholar Tom Moylan (2000, 196).

How do the novels compare as critical dystopias? In *Oryx and Crake*, one detects only a glimmer of hope at the end of the novel as Snowman discovers other human survivors. *The Year of the Flood*, however, offers a stronger gesture towards some form of emanci-pation for humanity. In his review of Atwood's second MaddAddam book in the *London Review of Books*, Marxist literary critic Fredric Jameson points to the novel's 'forms of resistance … [that] range from the survival of the most sadistic to the banding together of small groups and the formation of new religions or, more ominously, to what is called 'bioform resistance'' (2009, 7). Here, Jameson is referring to the God's Gardeners, whose enigmatic leader Adam One is configured as the messiah. This character is collected and strategic but just as equally mysterious and mystical. He represents logic and superstition at once.

Glorifying aspects of 'secular' ideas and personages through Christian rhetoric, Adam One's doctrine of salvation must not be simply described as an ecotheology, but as an eco-*teleology* given its messianic disposition. In a sermon commemorating the Festival of Arks, Adam One alludes to the biblical tale of Noah and describes an impending disaster that the Gardeners have come to call the 'Waterless Flood':

> Let us today remember Noah, the chosen caregiver of the Species. We God's Gardeners are a plural Noah: we too have been called, we too forewarned. We can feel the symptoms of coming disaster as a doctor feels a sick man's pulse. We must be ready for the time when those who have broken trust with the Animals – yes, wiped them from the face of the Earth where God placed them – will be swept away by the Waterless Flood, which will be carried on the wings of God's dark Angels that fly by night, and in airplanes and helicopters and bullet trains, and on transport and other such conveyances. (Atwood 2010, 91)

A nuanced difference can be delineated between Noah's disaster and the Gardener's Waterless Flood. To begin, the latter is not *quite* as prophetic as the former if we consider that Crake had secret dealings with the God's Gardeners in the past when he was better known as Glenn, smuggling Pilar's biopsy samples in a jar of honey to the HealthWyzer labs for diagnosis (Atwood 2010, 178). This alludes to the possibility of a conspiracy between Crake and Adam One in bringing about the Waterless Flood.

A more notable difference between the two relates to content. If one were to scour the Bible, one would stumble upon verses suggesting that God flooded the earth during Noah's time because humanity's 'wickedness…was great' and their conduct was 'evil continually' (Genesis 6:5). Adam One, however, is more explicit about the genesis of the Waterless Flood, blaming it on 'those who have broken trust with the Animals', thus upsetting the natural order of things. The biblical narrative of Noah's disaster accords God a greater role in the destruction of the earth whereas Adam One's sermon sees humanity as being solely responsible for its own unravelling. Indeed, the Waterless Flood will be delivered by man-made creations such as airplanes, helicopters and bullet trains. Seen this way, the Waterless Flood as an earth-shattering disaster gestures to 'overhumanisation' in a manner much more pronounced than its preceding myth.

The eco-teleology of the God's Gardeners strips the onus of agency from a divine being, putting it squarely onto humanity. It references John D. Caputo's 'weakness of God' theology, in which 'rethinking God on the model of something unconditional without power' (2006, 33) makes us imagine sovereignty not in terms of the possessives 'ours' or 'theirs'. Rather, the idea of a God *sans* power pushes us to envisage 'a more radical notion of democracy, in which each and every person, each and every thing, is included in the process' (Caputo 2006, 33). For Caputo, the weakness of God is not so much an admission of frailty, but a proclamation of might precisely because it is a voluntary act. It is weakness that draws strength from hyperbole: 'The weakness of God is *greater* than the strength of man; the foolishness of God is wiser than the foolishness of men' (Caputo 2006, 29). This sense of all-inclusive sovereignty is an ecological gesture because it begins within humanity a process of anti-anthropocentrism. Just as God sheds His power in the wider universe, so too should 'Man' [*sic*] strip himself of authority in the terrestrial world. No longer the apex of creation, humanity is now one among the flora and fauna, and not the boss of them, as mentioned earlier. Verses from the prayer 'Oh Let Me

Not be Proud', taken from *The God's Gardeners Oral Hymnbook*, lends credence to this impulse:

> Oh let me not be proud, dear Lord,
> Nor ran myself above,
> The greater Primates, through whose genes
> We grew into your Love. . . .
> And if we vaunt and puff ourselves
> With vanity and pride,
> Recall Australopithecus,
> Our Animal inside. . . . (Atwood 2010, 54)

As supplication, the above urges restraint against 'overhumanisation'. Its reference to the extinct bipedal primate species, Australopithecus, is meant to induce a sense of humility among those who encounter it. Scientists have speculated that the Australopithecus is an ancestor of the modern human, though far less intelligent. Meanwhile, the post-human being of the MaddAddam novels as embodied by the Crakers is not intellectually superior to the modern human, even if their bodies are better equipped to survive with less. On the contrary, the Crakers possess a childlike innocence that makes them somewhat subservient to the whims and fancies of the only modern human being they know, Jimmy/Snowman. As a case in point, the Janus-faced protagonist of *Oryx and Crake* is able to manipulate them into serving him a fish every week not just as a gesture of gratitude in exchange for a story but more so as a symbol of something divine. Jimmy's fish expedition has in fact been appropriated as a religious ritual in which the route between the fishing pond and the Crakers' encampment became christened the 'Snowman Fish Path'. One can almost see these expeditions as pilgrimages.

Yet one must be careful not to misread this messianic ideal embodied through Adam One and God's Gardeners in *The Year of the Flood* as signalling Atwood's enthusiasm about religion. To reiterate an earlier point, the texts are also suspicious of organised religion if we consider that the liobam was conceptualised by religious fanatics and not secular scientists. Furthermore, the example of the Crakers also lends credence to Atwood's somewhat reticent stance on religion. These blue pseudo-humans are not just post-human but also pre-human if one considers how their crude ideas about religious belief mirror the behaviour of Australopithecus and Neanderthals.

Other than its messianic disposition, one can outline yet another distinctive feature of Atwood's eco-teleology in *Oryx and Crake* and *The Year of the Flood* couched in the premise that eco-teleology can only be expressed *antithetically* through the imaginative genre of dystopian literature, thereby suggesting that this antidote to humanity's overhumanisation drive can only be figured as a form of negative theology. That is to say, faith's role as humanity's saviour against the apocalypse is affirmed through negation or the process of apophasis. This is first and foremost a hermeneutical venture, wherein the act of meaning-making is problematised and democratised so as to challenge an entrenched doctrine. To illustrate this is to revisit Adam One's sermon commemorating the Festival of Arks. The following excerpt from *The Year of the Flood* is instructive:

> 'And the fear of you' – that is, Man – 'and the dread of you shall be upon every beast of the earth, and upon every fowl of the air …into your hand are they delivered.' Genesis 9:2. This is not God telling Man that he has a right to destroy all the Animals, as some claim. Instead it is a warning to God's beloved Creatures: *Beware of Man, and of his evil heart.* (Atwood 2010, 90–91)

If the above can be taken as saying something about the nature of divinity, it is that the sacred is affected by the politics of translation. Adam One argues against a normative reading, against a tradition even, which posits that the God of Christianity has wanted only humans to understand His Message. Adam One re-translates that line from Genesis against the lens of anthropocentrism by suggesting that humanity's 'Other', the non-Man, or the Animals as they are referred to above, are just as literate. Only by negation, or the denial of a received doctrine, can eco-teleology make sense to the God's Gardeners. The protection of animals and other non-Man beings from overhumanisation necessitates a radical act of disavowal, the assertion of what Man is not – he is *not* benign, *nor* superior as the normative interpretation would have it. Humanity, as Adam One asserts by apophasis, is divinely ordained with an 'evil heart'.

At this juncture, more needs to be said about the renunciation of scientism in the MaddAddam novels. This is apophatic because it does not simply reject Crake's post-human vision of a bioengineered world inhabited by a self-sufficient pseudo-human species. Rather, the novels offer an alternative vision in which the hope for a better world is not invested in inventions but the ability to think creatively. The meaning of 'utopia' is not denied and emptied, but differed and deferred in the way that Derrida defines *différance*. That is, the future necessitates the absence of science because it signals the presence of the post-secular as expressed through the postulation that humanity's propensity for imagination could not be eradicated. In *Oryx and Crake,* Snowman reaches an epiphany:

> They understood about dreaming, he knew that: they dreamed themselves. Crake hadn't been able to eliminate dreams. *We're hardwired for dreams*, he said. He couldn't get rid of singing either. *We're hardwired for singing.* Singing and dreams are entwined. (Atwood 2004, 352)

As the next step in human evolution, the Crakers are imagined as imaginative, thus suggesting that the post-human condition is governed by the double appeal to utilitarianism and utopianism. That is to say, Humanity 2.0 is corporeally more adjusted to tackle a harsher climate but just as well predisposed to speculate 'other worlds'. Viewed from the perspective of 'natural selection', imagination does not merely serve 'a biological purpose' (Atwood 2004, 167) as Crake posits in *Oryx and Crake*, intimating that purpose to be an enabler to mating in that the more artistic a person is, the more likely s/he is able to secure a partner for copulation. Rather, imagination resides at the core of human survivability. It has enabled the Crakers to construct narratives, perform rituals and ultimately theorise a 'religion' to deal with uncertainties that might incapacitate them as a community, such as the fear of Snowman not returning from his trip, as well as the very mystery of the purpose of their existence.

In denying scientism, Atwood's apophatic eco-teleology is also reticent about the postulation that ecology is brimming with raw, boundless energy. In *Oryx and Crake*, Crake expresses cynicism at the idea of Nature 'with a capital N' when he tells Jimmy that 'Nature is to zoos as God is to churches' (Atwood 2004, 206) as he introduces the latter to the wolvogs, a creature engineered to look like a harmless dog but behaved like a ferocious pit bull. To Crake, human agency is the only legitimate force in the MaddAddam world, one that has tamed faith and ecology through technology. That is to say, biological advances have allowed humans to breathe life into creatures such as the wolvogs and pigoons, thereby casting humanity – not God – in the role of the Creator.

Yet the post-apocalyptic MaddAddam world suggests that this anthropocentric hubris is not sustainable. Both texts are filled with instances in which the human subject is shown to succumb to the might of religion and Nature. With religion, this essay has highlighted the potency of the belief in a higher power through the examples of the Crakers and the God's Gardeners. When it comes to Nature, this is intimated through the ways in which man-made creatures such as the pigoons easily establish an ecosystem following the Waterless Flood, becoming the apex predator of a newfangled food chain in which humans have become one of several preys. Such is the case with the aforementioned pigoons' hunt for Snowman in *Oryx and Crake*. A fuller scene was depicted in *The Year of the Flood* through the protagonist Toby as she observes the post-apocalyptic world beyond the confines of the Spa Café where she has fortified herself to survive the outbreak:

> Toby turns her binoculars skyward, at the crows racketing around. When she looks back, two liobams are crossing the meadow. A male, a female, strolling along as if they own the place. They stop at the boar, sniff briefly. Then they continue their walk. . . .
> Still, the liobams seem gentle enough, with their curly golden hair and twirling tails. They're nibbling flower heads, they don't look up; yet she has the sense that they're perfectly aware of her. Then the male opens its mouth, displaying its long, sharp canines, and calls. It's an off combination of baa and roar: a bloar, thinks Toby. (Atwood 2010, 94)

Designed as a splice between the lion and the lamb, the liobam resembles very little of the existing natural world outside of the MaddAddam novels. This sense of abnormality is intimated through Toby's reaction ('Am I imagining things?') as she encounters them. Yet the above scene also plays out like your run-of-the-mill *Discovery Channel* documentary narrated by the British naturalist David Attenborough. The liobam may appear bizarre, but its behaviour as the apex predator is naturalistic. The MaddAddam novels intuit poetic justice in recognising that man-made creatures have ironically made humans a food source. In the novels, the idea of nature as a stable, legible and controllable phenomenon is destabilised. The novels instead outline a wild post-human ecology where the 'winner' – be it humans, pigoons or Crakers – is unpredictable.

'Not-Yet' humans

As much as the texts can be said to be apophatic because they engage in acts of perennial denial (or, *différance*) when viewed as eco-teleology, they can also be described as constructing a non-polemical view of the human subject as that which occupies the liminal state between the terrestrial and otherworldly spheres. This view of humanity as a 'thing in between' is an integral aspect of the doctrine of 'theological humanism' developed by Klemm and Schweiker who write:

> The human is a 'bridge' (to speak metaphorically) between realms of life, at once animal and yet exceeding our animality. Human beings, in more biblical terms, are dust that breathes, made of the earth and yet an image of God. The decisive question [. . .] is whether 'humanity' is an *origin* or a *destination* or, we now add, lived in the *tension* of both. (2008, 17)

The above implies that the human condition is premised on incompleteness, being neither the beginning nor the end. However, this is not necessarily a cursed existence. As beings who are constantly vying to 'exceed our animality' in order to reclaim what was divine, humans are invested with a desire to become whole. Klemm and Schweiker cite the observations of the Hungarian philosopher Laszlo Versényi who describes humanity as 'a movement, a transcendence' (2008, 17).

In *The Year of the Flood*, this tension between the terrestrial and heavenly, the beast and the angel, takes centre stage with the God's Gardeners. On the one hand, the group's theology is littered with reminders of 'our Animal inside' to cite the 'Oh Let Me Not Be Proud' prayer from the *God's Gardeners Oral Hymnbook*. Indeed, the above-mentioned reference to the 'Australopithecus' serves to warn humanity against haughtiness. The same can be said of the 'The Feast of Adam and All Primates' sermon, in which Adam One speaks of humanity as having 'betrayed the trust of the Animals, and defiled our sacred task of stewardship' (Atwood 2010, 52–53).

On the other hand, the group's canonisation of Dian Fossey and Terry Fox is more than an exercise of desecularisation. It can also be seen as an attempt at transcending a degraded present. Both figures fought against insurmountable odds to advance what they believed are causes bigger than themselves. The naturalist Fossey, for instance, was renowned for her principled stand against poaching and tourism in Rwanda. Found murdered under mysterious circumstances in her remote cabin in the Virunga Mountains of Rwanda in 1985, the last entry in her diary is indicative of her utopian impulse, her desire for a better world: 'When you realise the value of all life, you dwell less on what is past and concentrate more on the preservation of the future'. This same sense of striving for a better future fuels Adam One's beatification of Terry Fox. Despite having one of his legs amputated for bone cancer, Fox embarked on a cross-country run across Canada in 1980 to raise awareness and funds for cancer research, but unfortunately died before he could complete his run as the cancer spread to his lungs. In his 'Saint Terry and All Wayfarers' sermon, Adam One speaks admirably of Fox's effort to transcend his corporeal limitations:

> Saint Terry's Day is dedicated to all Wayfarers – prime among them Saint Terry Fox, who ran so far with one mortal and one metallic leg; who showed what the human body can do in the way of locomotion without fossil fuels; who raced against Mortality, and in the end, outran his own Death, and lives on in Memory. (Atwood 2010, 403)

Appropriated as a saint, the struggle of Fox is immortalised as myth, or as 'Memory', by the God's Gardeners. In describing his tenacity, the sect has not only transformed Fox into a saint but also suggested that he is superhuman, almost angelic in disposition. Postulated as a 'thing in between', the personage of Fox is rhetorically leaning on the side of divinity rather than animality.

While other examples of this tension abound in the MaddAddam texts, it is apt at this juncture to juxtapose the thoughts of Ernst Bloch with Klemm and Schweiker's. Like the latter pair, Bloch inscribes emancipation into the idea of human incompleteness. In his voluminous magnum opus *The Principle of Hope*, Bloch theorises the 'Not-Yet' (*noch nicht*) as being central to the human condition, emphasising its transcendental nature. Like Atwood, Bloch argues for an apophatic doctrine of emancipation by focusing on the 'Not' of this 'Not-Yet' condition. He writes:

> The Not is lack of Something and also escape from this lack; thus it is a driving towards what is missing. Thus the driving in living things is depicted with Not: as drive, need, striving and primarily as hunger. (1996, 306)

Thus for Bloch, the 'Not' of the 'Not-Yet' condition is defined by a dissatisfaction with the here and now, a recognition that the present is degraded. This plays out in *Oryx and Crake* through Crake who saw his pre-apocalyptic MaddAddam world as decadent, incomplete, an interregnum in human evolution. With his utopian impulse evoked, Crake embarks on a mission to correct what he saw as a natural disorder, taking it upon himself to 'complete' the process of human evolution. Nature, with a capital N, becomes subservient to humanity, or so claims Crake. The novels are too dark to be considered literary utopias but their sense of measured hope brimming underneath a dystopian setting qualifies them as 'critical dystopias', as this essay has illustrated. It is important to note that critical dystopias 'expose the horror of the present moment', according to Moylan (2000, 169). In the MaddAddam world, that horror is not simply Crake's dissatisfaction with the here and now. Rather, it is the overhumanisation brought about by scientism that, left unchecked, could possibly result in the annihilation of humanity. Even if the texts make the hopeful gesture of life after apocalypse in the form of the Crakers, this is itself mired in incompleteness. Thus Crake's post-human solution becomes a Blochian utopian impulse – out of Crake's sense of the 'Not-Yet' future was born 'Not-Yet' humans. Despite Crake's best effort to eliminate abstract thinking among his blue-skinned pseudo-human creation, the Crakers have developed a proclivity to practice religion and art. The superhuman has been humanised, cursed with an unquenchable desire for something more. In postulating that *Oryx and Crake* and *The Year of the Flood* present a hopeful vision amidst the seeming all-engulfing hopelessness that punctuates the MaddAddam world, this essay agrees with Canavan's reading of the novels as utopian because they acknowledge that humanity is plagued by 'the strident insistence that things might yet be otherwise' (2012, 156). However, if this insistence or impulse is to be viewed as 'a radical break' (2012, 156) as Canavan describes it, then it must be one that is given to a process of negation. The hope for a better world is *not* to be found within the certitude of Science, but the perennial discontentment of the humanly 'Not-Yet' condition.

References

Atwood, M. 2004. 2003. *Oryx and Crake*. New York: Random House.

Atwood, M. 2010. 2009. *The Year of the Flood*. New York: Random House.

Atwood, M. 2011. "The Road to Ustopia." *The Guardian*, October 14. Accessed July 9, 2013. http://www.guardian.co.uk/books/2011/oct/14/margaret-atwood-road-to-ustopia

Baccolini, R. 2004. "The Persistence of Hope in Dystopian Science Fiction." *PMLA* 119 (3): 518–521.

Bloch, E. 1996. 1959. *The Principle of Hope: Volume 1*, edited by N. Plaice, S. Plaice, and P. Knight. Cambridge, MA: MIT Press.

Canavan, G. 2012. "Hope, But Not for Us: Ecological Science Fiction and the End of the World in Margaret Atwood's *Oryx and Crake* and *the Year of the Flood*." *Lit: Literature Interpretation Theory* 23 (2): 138–159.

Caputo, J. D. 2006. "Beyond Sovereignty: Many Nations Under the Weakness of God." *Soundings: An Interdisciplinary Journal* 89 (1/2): 21–35.

Curry, P. 2011. *Ecological Ethics: An Introduction*. 2nd ed. Malden, MA: Polity.

Jameson, F. 2009. "Then You Are Them." *London Review of Books* 31 (17): 7–8.

Klemm, D. E., and W. Schweiker. 2008. *Religion and the Human Future: An Essay on Theological Humanism*. Malden, MA: Wily-Blackwell.

Moylan, T. 2000. *Scraps of the Untainted Sky: Science Fiction, Utopia, Dystopia*. Boulder, CO: Westview-Perseus.

Sorell, T. 1991. *Scientism: Philosophy and the Infatuation with Science*. London: Routledge.

Genre, utopia, and ecological crisis: world-multiplication in Le Guin's fantasy

Katherine Buse

Faculty of English, Cambridge University, Cambridge, UK

How should the utopian potential of a literary text be identified or evaluated? Speculative fiction is often read in terms of its rendering of alternative worlds that seem 'possible.' Ecological crisis, however, even as it raises the question of how a livable world can be imagined, also breaks down the sense that a 'better' world is possible. This article examines a debate over utopia between Darko Suvin, whose literary criticism emphasises the rational and cognitive aspects of SF, and author Ursula Le Guin, who critiques Suvin's 'Euclidean' definition of utopia. Le Guin's novel *The Farthest Shore* illustrates her alternative definition of literary utopia, suggesting that in times of ecological crisis, seeking utopia must transcend the boundaries of 'the possible.' Using the genre of fantasy and its resonances with Rachel Carson's depiction of ecological crisis, Le Guin highlights the function of narrative and naming in bringing about a livable world.

Ursula K. Le Guin's novel *The Farthest Shore* opens with reports of a developing ecological crisis. A farming community has been beset by inexplicable hardship: 'there was sickness among them, and their autumn harvest had been poor,' and the problem appears to be spreading across the archipelago in which the novel is set (1993, 305). Arren, the diplomatic envoy, tells of 'trouble among the flocks this spring, the ewes dying in birth, and many lambs born dead, and some are... deformed. [...] "I saw some of them," he said. There was a pause' (1993, 306).[1] The imagery, which references that particularly ovine form of agriculture most associated with the pastoral, evokes notions of a bucolic springtime uncannily disrupted. The textual hesitations recall the pacing of cinematic suspense, conveying an affect of horrified realisation, lurking and unarticulated significance.

Poor harvests, dying greenery, deformities and other ailments are characteristic of what Lawrence Buell labelled 'toxic discourse.' In his essay by the same name, he remarks upon Rachel Carson's *Silent Spring* as the first modern appearance of a now-ubiquitous discursive aesthetic: 'birth defects and miscarriages among families,' fields and neighbourhood yards 'that ought to be green [...] but instead show only sparse, half-dead plant cover [...] unnatural-looking chemical soup' (1998, 646). In 'A Fable for Tomorrow,' the preface to Carson's book, a small community's pastoral life is disrupted: Carson writes of a 'strange blight,' which 'crept over the area and everything began to change,' of 'mysterious maladies' afflicting livestock and people and leaving townsfolk 'puzzled and disturbed' (1962, 13). *Silent Spring* is non-fiction, but it operates on a mythic level, producing the sense of 'an awakening [...] a horrified realization that there is no protective environmental blanket, leaving one feeling dreadfully wronged' (Buell 1998, 646). If Carson's text is the cornerstone of toxic discourse in the contemporary age, *The Farthest Shore* seems to replicate many of its images, themes, and philosophical questions. Both begin with a crisis in the pastoral, both take as their conflict a scorched-earth campaign of ecological destruction motivated by human greed, both meditate on the role of human beings in the non-human world, and both present a (metaphorically) 'silent' spring. The resonances between the two texts are striking, and hardly a coincidence: Le Guin's novel was published in 1972, a decade after *Silent Spring*, and (as evidenced by her contributions to the *Whole Earth Catalogue*, for example) Le Guin was avidly engaged in the modern environmental movement that claims *Silent Spring* as a foundational text.

However, what may be less immediately obvious is the extent to which both texts claim a common ancestry. Le Guin is not merely evoking Carson's landscapes of toxicity, she is writing the third fantasy novel in her Earthsea sequence – and Carson herself drew upon the resources of fantasy to produce the affects for which *Silent Spring* is famous, and to which it owes a share of its success. Carson's opening parable clearly draws upon the fantastic and the mythic, describing pollution in terms of an 'evil spell,' 'witchcraft,' and a 'grim spectre' (1962, 13–14). But even in the chapters that are fully non-fictional, Carson continues to refer to sorcerers' robes, alchemy, and witchcraft. When she describes 'a weird world, surpassing the imaginings of the brothers Grimm,' where, like 'the enchanted forest of the fairy tales,' a creature can die from 'vapors emanating from a plant it has never touched,' she is gaining currency from tropes developed in fantasy (1962, 39).

As will be discussed in greater depth, the conceptual and formal ties between fantasy and ecological crisis are intimate. While the connection between fantasy and the environment has been extensively explored in the field of SF studies, the growing number of ecocritics researching SF have on the whole focused on the parts of the tradition, such as 'hard SF,' which imagine naturalistically plausible events.[2] This is unsurprising, because many of these ecocritical explorations are driven by a concern with the future of our earthbound environment: ecological crisis demands the adaptation (and sometimes abandonment) of ecocritical approaches that traditionally relied on ideas like 'the natural,' the picturesque, the local or the straightforwardly mimetic.[3] SF thus often appears in ecocriticism as a supplement to these traditions in response to ecological crisis' implication of unfamiliar geographical scales, time frames, and epistemologies.[4]

Insofar as the questions asked of SF are related to 'our' own future as perpetrators and victims of ecological harms, they serve a utopian function: in the present day, we are already 'too late' to prevent the decline of many of the environmental conditions that will sustain whatever future is imminent. How can a better world, or even a world that is not objectively 'worse,' emerge from such a moment? The pressures of ecological crisis have shaped ecocritics' expectations that SF texts be 'utopias of the possible,' such that the plausibility and desirability of the imagined SF world – as if it were a potential future – is a cornerstone of the selection and analysis of texts. For example, an introduction to an ecocritical special issue of *Ecozon@* suggests that the most characteristic SF texts are either 'ask[ing] by their very structure how we might achieve their utopias or prevent their disasters and dystopias' or 'seek[ing] to persuade the reader of the current reality of ecological crisis' (Heise 2012b, 4–5). The tendency to focus upon the 'possibility' of SF (its perceived ability to offer concrete insights about the physical world) seems to derive from the pairing of concerns about future ecological changes with an ecocritical tradition in which 'success is usually decided by the yardstick of mimesis' (Johns-Putra 2010, 757). SF is non-realist, but when realism no longer applies, it may seem to be the next-best thing: Adeline Johns-Putra writes that in the case of climate change, authors and critics 'are *compelled* to draw on the strategies of one of the primary genres of futuristic imagining: science fiction' (2010, 749, my emphasis). It is perhaps because of this reluctance to stray too far from mimesis that a striking majority of ecocritical interventions into SF cite Darko Suvin,[5] a critic who famously defined SF as the literature of 'cognitive estrangement,' and whose criticism heavily relies on notions of 'the possible.'

SF offers opportunities for a range of literary-critical methodologies and a rich terrain for exploring how societies think about ecology and ecological crisis.[6] In order to demonstrate the merits of expanding ecocritical approaches to SF, this analysis traces a debate about utopia between Suvin and Le Guin. While Le Guin has written several well-known science fiction utopias, such as *The Dispossessed* and *Always Coming Home*, it is her genre fantasy sequence, the *Earthsea* trilogy, and especially its third novel *The Farthest Shore,* which most compellingly addresses the question of imagining utopia in environmental crisis. While the world of *Earthsea* is entirely *im*possible, Le Guin's use (and variation) of the fantasy genre's conventions shows how generic expectations shape real-world understandings and how non-mimetic texts nonetheless offer insights into contemporary scenarios. In my analysis, Le Guin takes advantage of Carson's iconic 'toxic discourse' and its debt to genre fantasy to re-evoke the scenario limned by such a discourse. Le Guin goes on to revise Carson's text, crafting a fantasy narrative's closure to Carson's open-ended 'Fable for tomorrow.' She depicts this as a battle over how to find 'the good place,' in order to make her own argument about why the 'Euclidean [Suvinian] utopia' seems to break down in a future marked by environmental crisis, and what her own model of utopia has to offer as an alternative. Le Guin's position suggests that the domain of literature and language – while it may speculate upon technological plausibilities – has its most significant utopian role less in the building of a 'new' or 'possible' world than in structuring ways of experiencing, engaging, reading, and interpreting any world that is to be worth living in.

Earthsea and magic

Le Guin's original Earthsea trilogy comprises what are arguably her most beloved pieces of fiction (Clute and Nicholls 2012, 'U. K. Le Guin'). The trilogy follows the career of a wizard, Ged, through his childhood coming of age, his heroic exploits in adulthood, and as an ageing Archmage, the head of the school that teaches magic on the islands. In Earthsea, the mechanism of magic is a language, True Speech. As Le Guin describes it, 'the essence of the art-magic as practiced in Earthsea' is naming, and 'to know the name of an island or character is to know the island or the person' (1979, 41). True Speech is always true, and so it can be used to make changes in the real world. It is explicitly ecological, not only in that it describes a natural world, but in Le Guin's evocation of its interconnectedness and embeddedness in the universe's ecology: 'All power is one in source and end [...] years and distances, stars and candles, water and wind and wizardry, the craft in a man's hand and the wisdom in a tree's root: they all arise together,' Ged says in the trilogy's first novel (1968, 185). The mysticism of this passage nearly obscures a sly reference to Einsteinian relativity (an important aspect of Le Guin's science fiction utopias, such as *The Dispossessed*), and intertwines a vastness both vertical and lateral, depicting forces and energy which are co-emergent not only with ecology but human language, material culture, and non-human nature. Ged continues, 'My name, and yours, and the true name of the sun, or a spring of water, or an unborn child, all are syllables of the great word that is very slowly spoken by the shining of the stars' (1968, 185). True Speech is a way of speaking the language of the universe, and it is enabled by the fact that everything has a name and is significant as part of a larger expression, 'a great word [...] spoken by the shining of the stars' (1968, 185).

Utopias of the possible

It is precisely because of True Speech that Darko Suvin, a friend of Le Guin's, rejects Earthsea as escapist: it 'seems to [...] conflate unduly poetry [...] and political economics' (2006, 501). Suvin's definition of SF is by far the most exclusive of 'unscientific' texts since the notion that SF would actually predict the future fell out of fashion in the 1950s.[7] This is because his revolutionary utopian framework wants SF to promise that another world – a better world – is not just fantasmatically but materially possible. For this reason, Suvin has drawn criticism for dismissing any potential value in a vast swathe of speculative fiction,[8] as in this now-infamous statement: 'Even less congenial to SF is the fantasy. [C]ommercial lumping of it into the same category as SF is [a] rampantly socio-pathological phenomenon' (1979, 8–9). For Suvin, 'science fiction' is 'both a descriptive and an evaluative term,' which is to say that 'bad sf is not sf' (Csicsery-Ronay 2003, 119). As for how to separate the good from the bad, Suvin is concerned less 'with the terms of art than with terms of knowledge of social truth' – what is at stake is a particular yearning for or understanding of utopia (Csicsery-Ronay 2003, 119).[9]

For Suvin, SF 'has always been wedded to a hope of finding in the unknown the ideal environment, tribe, state, intelligence, or other aspect of the Supreme Good,' and if it fails in this regard, at least 'the *possibility* of other strange, co-variant systems and semantic fields is assumed' (1979, 5, my emphasis). In his 1979 *Metamorphoses of Science Fiction*, Suvin defines SF as 'a literary genre whose necessary and sufficient conditions are the presence and interaction of estrangement and cognition' (7–8). Istvan Csicsery-Ronay's gloss of cognitive estrangement illuminates the ways in which cognitive estrangement separates science fiction from other genres: 'Science-fictional estrangement works like scientific modelling: the familiar [...] situation is either rationally extrapolated to reveal its hidden norms and premises [or] analogically displaced on to something unfamiliar [to reveal unnoticed] elements' (2003, 118). One might note that on its own, this description might not exclude fantasy at all. However, the 'specific difference between sf and other estranging genres, such as fantasy, is that sf's displacements must be logically consistent and methodical; in fact, they must be scientific to the extent that they imitate, reinforce and illuminate the process of scientific cognition' (118). Further rationalising Suvin's theory is 'the novum,' a specific object or change introduced by the author which is able to convincingly explain the transformation that leads to the fictional world's estrangement from our 'reality.' For Suvin, the closer an imagined world is to *actually possible*, based on contemporary scientific knowledge, the more likely it is to be utopian SF, both reflecting upon 'reality' and remaining consistent with the more abstract 'reality' of science. To find such a utopia is to differentiate between historical 'signal' and 'noise,' determining which elements of the real world are fixed, and which can change, as well as which variables are independent and dependent.

For Suvin, the intrusion of magic into an SF story marks it as a failure to imagine a 'better' world, and is a distraction from SF's utopian function. Unsurprisingly, therefore, while Suvin has frequently praised Le Guin's science fiction utopias, he has demonstrated very limited patience with her fantasy.[10] A footnote which appears in his essay, 'On U. K. Le Guin's 'Second Earthsea trilogy' and its cognitions,' contains a very personal public apology which thanks Le Guin for inspiring the piece by sending him 'an angry letter about my essay on Fantasy in *Extrapolation* [...], and in particular about what I had to say in it on Earthsea' (2006, 503). His penitence is perhaps because he has consistently used Earthsea as a representative of all fantasy, whilst criticising the genre for escapism, a 'general absence of cognitiveness [...] bound up with denial or repression of key elements of earthly history,' and thus for being useless for anything but the most 'indirect cognitive transfer' (2006, 499–500). However, Le Guin's True Speech, insofar as it relates to the idea of 'naming' the world, bears strong resemblances her writing about utopia in her non-fiction, and thus merits serious examination, even if it looks like non-cognitive and anti-utopian 'magic.'

Le Guin's utopia

Le Guinean naming is no Garden-of-Eden assignment of titles. As suggested by the notion that names are 'syllables' in the mechanics of Earthsea's universe, a name must be discovered (not invented). The implication of the idea that everything has a name is spelled out more concretely by one of his teachers when Ged is a young man: he learns that if a wizard were to attempt to 'lay a spell [...] over all the ocean,' he could not just say the word for 'sea,' but would have to know 'the name of every stretch and bit and part of the sea [...] and beyond to where names cease. [...] A mage can control only what is near him, what he can name exactly and wholly' (50). The first qualification of Le Guin's magic is that humanity's ability to make changes is not limitless, but constrained by a world in which everything has a name and cannot be controlled unless it has been attended to. Thus, 'what a wizard spends his life at is finding out the names of things' (267). While some wizards find names in books, Ged's teacher Ogion derives them himself by living with elements of the non-human world, wandering forests and refusing to cast spells to keep himself dry in the rain. In first describing the process of magic to a young Ged, his mentor corrects him when he asks about 'the use' of a particular herb:

> When you know the fourfoil in all its seasons root and leaf and flower, by sight and scent and seed, then you may learn its true name, knowing its being: which is more than its use. What, after all, is the use of you? or of myself? Is Gont Mountain useful, or the Open Sea? (26)

In order to name something, a wizard must spend time with it and discover its name through intimacy and careful attention. This means that the subject–object relationship suggested by the idea of 'using something' is subverted by the logic of Earthsea's magic. To know the name that allows magic to be performed is to know the 'being' of a thing, to have a relationship with it that renders it an end in itself.

Naming links Le Guin's notion of utopia and the world in which she sets Earthsea and its stories. In her non-fiction, such as her 1989 essay, 'A non-Euclidean view of California as a cold place to be,' Le Guin writes that theories of utopia have been 'trapped, like capitalism, industrialism, and the human population, in a one-way future consisting only of growth,' a critique which will be elaborated in greater depth later in this discussion (1989, 10). Le Guin expresses a hope that 'our final loss of faith in that radiant sandcastle may enable our eyes to adjust to a dimmer light and in it perceive another kind of utopia' (1989, 10). Namely, 'in order to speculate safely on an inhabitable future, perhaps we would do well to find [a] sense of where we live, where we are right here right now,' because if 'we did, we wouldn't muck it up the way we do. If we did, [...] we might have some sense of our future as a people. We might know where the center of the world is' (1989, 6). To explain this articulation of utopia, she writes about California Indian cultures for whom names are not merely generic or arbitrary words, but a way of imbuing the world with meaning:

> Every hill, every valley, creek, [...] beach, bend, goodsized boulder, and tree of any character had its name, its place in the order of things. An order was perceived, of which the [white] invaders were entirely ignorant. Each of those names named, not a goal, not a place to get to, but a place where one is: a center of the world. (1989, 10)

The notion of each name naming a 'centre of the world,' a sort of local and provisional utopia, has considerable biographical significance for Le Guin, as her parents, prominent anthropologists, 'studied this use of naming to order reality in the Yurok and other California Indians' (Rochelle 2001, 6). Cleverly, Le Guin collides Robert C. Elliott's definition in *The Shape of Utopia* – 'the utopia aspires to [...] a state of society in which human beings are at peace with themselves and their environment' – with historian Walton Bean's observation that 'the California Indians had made a successful adaptation to their environment and they had learned to live without destroying each other,' to suggest that utopia needs to be redefined not as progress towards eventual perfection, but finding a process worth persisting at as an end in itself (Elliot 1970, 100; Bean 1968, 4, both cited in Le Guin 1989).

The idea of the 'ecological Indian' has a long history in America, which has been problematised by work such as Shepard Krech's and has been debated extensively on the grounds of historical accuracy, ideological effectiveness, and Western psychology, to name but a few. However, Le Guin's own lines of influence are less reifying of otherness and more informed by her own unconventional upbringing, which may have put her in contact with ideas about language and story that diverge from a Western tendency to separate the 'real' from the 'unreal.' In this essay, Le Guin cites her father's *Handbook of the Indians of California,* which describes the Yurok and the Karok tribes (among others) as believing that articulating certain words, names, phrases, or narratives can produce immediate effects – Kroeber calls them 'magical.'

Therefore, while Earthsea features magic, there is no reason to presume that (as Suvin says) 'SF retrogressing into fairy tale [...] is committing creative suicide' by presenting a 'world indifferent to cognitive possibilities' (1979, 8). As I suggested earlier, the Suvinian paradigm leaves a mark on how texts are read, and it is perhaps a 'scientific'-seeming relationship to the notion of 'the possible' that initially attracted ecocritics to SF. In one essay using 'science fiction [...] to explore environmental futures,' (2012a, 99), Ursula Heise uses the word 'literal' not once but twice as she calls for 'environmentally oriented literary criticism' to heed the '*literal* ecological significance' of imagined worlds' ecologies (2012, 101, my emphasis). Likewise, Patrick D. Murphy complains that in Bruce Sterling's future history of climate change, 'education [about] likely effects of global warming [...] often gets overshadowed by attention to the [...] gadgetry' (2001, 239). SF is treated in such discussions as a space for *information* and/or realistic thought experiment about the ethical and eco-political considerations of our own future on earth, as if 'science fiction' were a more narrative version of 'science.' However, there is no obvious reason why SF must be evaluated solely in these terms. It is still a literary – rather than a literal – genre, and it responds to the tools of literary criticism at least as well as to examinations of its factual content.

Silent Spring, ecological crisis, and utopias of the possible

Le Guin's own writing on utopia strongly critiques of utopias of the possible, partly because of what she sees as the folly – and, often, unintended consequences – of attempting to realise a 'rational utopia,' or interpret it as if it were really possible:

> [I]t is of the very essence of the rational or Jovian utopia that it is not here and not now. It is made by the reaction of will and reason against, away from, the here and now, and it is, as More said in naming it, nowhere. (1989, 2)

Le Guin uses the language of mathematics to describe the characteristics that lead Suvin to call a utopia 'possible': 'Utopia has been euclidean, it has been European, and it has been masculine,' she writes (11). But after a history of failed utopian projects, Le Guin argues that it is no longer possible to imagine understanding, controlling, and sculpting a better world. The will and reason calculates a world that does not exist and never has, a 'pure structure without content; pure model,' and while the 'goal [...] is its virtue, [...] Utopia is uninhabitable. As soon as we reach it, it ceases to be utopia' (2). For Le Guin, such utopias must not be seen as 'possibilities' or 'predictions,' but rather as abstract concepts never to be materialised. The imagery of mathematical limits highlights the abstraction of the 'rational' approach to utopia and resonates with historical examples of hubristic and misguided techno-scientific projects.

While the human track record of ill-fated technological projects stretches at least as far back as stories of Babel and of Icarus, one of the most compelling examples in the late twentieth century is 'the contamination of air, earth, rivers, and sea with dangerous and even lethal materials' first brought to popular attention by Rachel Carson, a result of our species having 'acquired significant power to alter the nature of his world,' which reached 'disturbing magnitude but [also] changed in character,' as now the 'pollution is for the most part irrecoverable; the chain of evil it initiates [is] irreversible' (1962, 16). Expressing (as does Le Guin) a sense of the ultimate failure of 'rational' technoscientific projects, Carson writes, 'future historians may well be amazed [that] intelligent beings [sought] to control a few unwanted species by a method that contaminated the entire environment and brought the threat of disease and death even to their own kind [but] this is precisely what we have done' (19).

The remainder of this essay analyses how Le Guin engages with Carson's narrative, using the elements of the original – and its resonances with the fantasy genre – to comment upon the relationship between utopias of the possible and the causes and potential remedies of ecological crisis. Le Guin criticises Euclidean utopia's 'Models, plans, blueprints, wiring diagrams' as being driven by a desire 'to have power over what happens, to control. Knowledge is power, and we want to know what comes next, we want it all mapped out' (1989, 18). Carson, too, comments on her fears about a desire for power over reality, stating that 'chemical weed killers [...] work in a spectacular way; they give a giddy sense of power over nature to those who wield them' (1962, 69). Ged tells Arren, 'when we crave power over life [and] knowledge allies itself to that greed, then [...] the balance of the world is swayed, and ruin weighs heavy in the scale' (333–334). The notion of knowledge allied to greed maps onto the problem Carson limns in *Silent Spring*: the chemicals are 'the synthetic creations of man's inventive mind, brewed in his laboratories,' and this technoscientific mystery-meat is irresponsibly sold by companies operating in 'an era dominated by [...] the right to make a dollar at whatever cost' (17, 22–23).

In *The Farthest Shore*, Arren expresses puzzlement at the growing signs of environmental crisis: 'Can it be a kind of pestilence, a plague, that drifts from land to land, blighting the crops and the flocks and men's spirits?,' he asks Ged (333). This language of pestilence or blight 'drifting' strongly recalls *Silent Spring's* opening fable, which describes a 'strange blight' that 'crept [...] almost unnoticed,' as disease 'swept the flocks,' leaving communities 'disturbed' by the 'withered vegetation' and the land 'deserted by all living things' (1962, 12–14). Ged's response to Arren's question mirrors the parable's famous final line, in which Carson states that no foreign threat 'had silenced the rebirth of new life in this stricken world. The people had done it themselves' (1962, 14). Ged says, 'A pestilence is a motion of the great Balance, of the Equilibrium itself; this is different. [...] Nature is not unnatural. This is not a righting of the balance, but an upsetting of it. There is only one creature who can do that. [...] We men' (333). This echoes not only Carson's attribution of blame, but also the language that she uses to describe ecological systems, as when she writes of 'a natural system in perfect balance, [except] in those already vast and growing areas' which are 'entirely being unbalanced [by our] unnatural manipulations' (66, 261). Despite the very clear resonances between the two texts, as I suggested earlier, Le Guin's imitation of Carson does not stray from the fantasy genre – the language of blight, and balance, good and evil, had been characteristic of fantasy texts for nearly a decade when *Silent Spring* was written.

Fantasy and ecological crisis

Brian Attebery, in *Strategies of Fantasy,* (lovingly) satirises the contemporary fantasy tradition by presenting a recipe for a fantasy narrative.

> Take a vaguely medieval world. Add a problem, something more or less ecological, [and] one villain with no particular characteristics except a nearly all-powerful badness. [...] Pour in enough mythological creatures and nonhuman races to fill out. [To] the above mixture add one naive and ordinary hero who will prove to be the prophesied savior. [...] Keep stirring until the whole thing congeals. (1992, 10)

In addition to providing a strikingly accurate gloss of the fantasy formula, what this passage reveals is that, amongst those who engage with fantasy literature regularly (for literary critics, that is to say, a relatively isolated island of scholarship), it is well known that when there is a crisis in fantasy, it is 'something more or less ecological.' As one SF critic writes, if science fiction is driven by a desire for 'understanding,' then fantasy's 'motive force is loss [...] as if to say that the world is *wrong* and the story is to explore ways of putting it right' again (Sawyer 2006, 405). The most iconic modern fantasy narrative (which in Attebery's estimation is the common ancestor of the genre's contemporary form) directly references environmental crisis: 'In *The Lord of the Rings* [...] the wounding of the land specifically parallels the effect of the Industrial Revolution – slagheaps, polluted air and water, [– m]uch genre fantasy simply copies this' (Clute 1999f, 'Dark Lord'). Thus, although science fiction offers visions of 'the possible,' it is fantasy that (in its modern form) has been developed from the beginning as a response to, or commentary on, eco-crisis.

By describing what connects texts to one another, and 'dictating which textual details come into focus, [genre] brings with it its own understanding,' ways of extracting meaning from a text (Philmus 2005, 29). As with SF, the definition of fantasy is debated. But most cite John Clute's work in the Encyclopedia of Fantasy (1999), not so much for its definition as for its description of how fantasy is structured as a four-stage narrative progression.[11] Within this structure, the fantasy genre's use of setting helps to explain why it has such deep ties to environmental crisis, and why *Silent Spring* draws upon its tropes. As Suvin is fond of pointing out, 'The world of a work of SF is not a priori intentionally oriented towards its protagonists,' and thus it 'shares with the dominant literature of our civilisation' the assumptions 'of modern science and philosophy' (1979, 11). In fantasy, however, 'circumstances around the hero are neither passive nor neutral' (1979, 11). In fantasy, environment and setting take on a different meaning than in mimetic fiction: 'every nook and cranny, every chasm and crag, every desert and fertile valley is potentially meaningful. And how a landscape is described in fantasy is what that landscape means': the state of a land (as healthy, wild, quaint, desertified, or what have you) 'both defines [...] and is defined by' the hero (Clute 1999e, 'Land').

Landscapes are neither indifferent nor incidental in fantasy, but an indication of the way the narrative is progressing, for 'literatures of the Fantastic positively glory in the fact that they present, and embody, Story-shaped worlds' (Clute 1999c, 'Recognition'). It is for this reason that fantasy's structure is so intertwined with ecological crisis: the most characteristic and 'structurally complete' fantasy narratives follow a trajectory which moves from wrongness (the first evidence of 'a profound and inherent illness in [...] the essence of things') to thinning (almost directly akin to ecological destruction, a fading, dying or enfeeblement of the world and its properties) to recognition and then healing, in which the story's resolution is discovered and implemented (Clute 1999d, 'Wrongness'). The hero must discover the cause of the degradation of the land and identify his (or her) role in fixing it – this is all at once a diagnosis of the problem with the landscape, a moral for the story, and a discovery of the hero's role in the world, because they are the same thing.

Silent Spring and/as fantasy

Rachel Carson's *Silent Spring* assembled a body of knowledge about the effects of chemical insecticides that had never previously been considered, leading to both legislative reform and mass popular interest in pollution. Without specifically mentioning genre fantasy, several critics have noticed the book's debts to non-realist and non-cognitive generic traditions. Greg Garrard describes its preface as 'fairy tale' which marries the two genres of pastoral and apocalypse (2012, 2). Similarly, Frederick Buell describes it as 'reviv[ing] a long-standing mythography of betrayed Edens,' and later ties these 'traumas of pastoral disruption' to SF ecocatastrophe narratives. The rhetoricians Killingsworth and Palmer, in their analysis of Carson's use of language agree with pastoral and apocalypse, adding prophecy, fable, and gothic 'magic gone awry' (1992, 66–67). By the time they also refer to Carson's innovation as a 'famous appropriation of the science fiction genre,' one is relieved at the explanation that the prologue 'experiments with the shifting and overlaying of genres' (1992, 66–67).

What Garrard describes as the apocalyptic narrative, in particular, clearly trades upon fantasy's structure: 'every element of the rural idyll is torn apart by some agent of change, the mystery of which is emphasized by the use of both natural and supernatural terminology' (1–2). In this transition from changeless pastoral harmony to mysterious and all-pervasive destruction, Carson is performing a literary move with which readers of fantasy are extremely familiar. What 'is normally signalled in fantasy by Wrongness' is that 'the world [...] is about to undergo a dangerous and painful Thinning of texture, a fading away of beingness' (Clute 1999b, 'Fantasy'). This description highlights a movement from the local to the systemic that is encoded in the structures of fantasy. While wrongness is simply the first indication, an intrusion of something that does not make sense or should not be, thinning is the way that the world is corroded pervasively by the threat. A concept like 'pollution' gains its valence from a similar movement, an extrapolation of uncanny or inexplicable events (wrongness) to an intuition about the state of the world as a whole: poisonous insects or even dying birds, taken individually, are not as pervasive a threat as they seem when amassed as a set of narratives that challenge collectively-held senses of 'normality.'

What this suggests is that Rachel Carson used the iconography and generic expectations of fantasy to make the world of *Silent Spring* seem story-shaped to her readers. By exploiting the movement from incomprehensible wrongness to the implication that thinning is imminent, she demands our response: what is 'apocalyptic' about Carson's text is the all-encompassing 'sense that the world as a whole has gone askew,' a sense that derives not only from the evidence but from the expectation, imported from fantasy, that this thinning is somehow 'about' us, that is, that the world is story-shaped (Clute 1999b, 'Fantasy').

Le Guin mirrors many of Carson's images of wrongness: ruined landscapes, human illness, and even animals behaving in uncanny ways, manifesting strange attractions, and gruesome suffering. One example is her depiction of the addicts of a drug, hazia, which Ged and Arren discover as they begin their quest. The afflictions of hazia bear strong similarities to the chemicals in *Silent Spring*. Le Guin writes of hazia, 'the stuff is poison. First there is a trembling, and later paralysis, and then death' (338). Carson describes the effect of the chemical Aldrin almost identically: 'Birds picked up in a dying condition showed the typical symptoms of insecticide poisoning – tremoring, loss of ability to fly, paralysis, convulsions' (1962, 87–88). Le Guin describes a catatonic hazia addict, whose hand 'made a jerky, circular motion in the air, as if she had quite forgotten about it and it was moved only by the repeated surging of a palsy or shaking in the muscles. The gesture was like […] a spell without meaning' (338). This reflects Carson's description of a baby exposed to the chemical endrin, who 'became […] unable to see or hear, subject to frequent muscular spasms, apparently completely cut off from contact with his surroundings' (35). This resonance between the two texts – being cut off, disconnected, unable to produce meaning – begins to reveal what Le Guin diagnoses as the nature of the thinning that attends Carson's imagery of toxic wrongness: a destruction not of the world but of the sense that human beings have a meaningful and comprehensible place in it.

Le Guin's silent spring

Echoing Carson's suggestion that 'the rebirth of new life' had been 'silenced,' Le Guin also metaphorises the notion of a silent spring. However, *The Farthest Shore*'s 'stricken world' relates strongly to the silencing of language itself. Words in True Speech are losing their meaning, suggesting a slippage of the utopian principle of naming. As I discussed earlier, in order to name something, characters are supposed to be able to develop a certain relationship with it: of inhabitation, understanding, and intimacy. Le Guin depicts this ability to name things as threatened by ecological crisis: throughout *The Farthest Shore* there is a repeated motif that describes Arren as 'lost' in some incomprehensible space.

> There was horror in the earth and in the thick air, an enormity of horror. This place was fear, was fear itself; and he was in it, and there were no paths. He must find the way, but there were no paths, and he was tiny, like a child, like an ant, and the place was huge, endless. (375)

Like Arren, *Silent Spring*'s readers are presented with a vision of the earth that is huge, endless, and resistant to easy mapping: ecology is 'a complex, precise, and highly integrated system of relationships between living things' and despite Carson's reassurance that 'Man [sic], too, is part of this balance' (1962, 218), she seems to envision a fairly limited role: 'with a minimum of help and a maximum of noninterference from man,' she writes, 'Nature can have her way, setting up all that wonderful and intricate system of checks and balances that protects the forest from undue damage' (258). It seems that the only specific place in the world she can imagine for human beings is as antagonist – as if we were villains of nearly all-powerful badness, bent on wreaking 'undue damage' and preventing nature from having 'her' way.

> It is human nature to shrug off what may seem to us a vague threat of future disaster. [But for] each of us [...] this is a problem of ecology, of interrelationships, of interdependence. We poison the caddis flies in a stream and the salmon runs dwindle and die. We poison the gnats in a lake and the poison travels from link to link of the food chain and soon the birds of the lake margins become its victims. We spray our elms and the following springs are silent of robin song[. This is] the web of life – or death – that scientists know as ecology. (169–170)

Throughout this pattern of 'we poison,' repeated over and over, Carson's catechism seems clear: animal relationships are characterised by interdependence, creating a beautiful web of life. But 'human nature' – and the use of this term is not meaningless – is rather to spread a 'web of death.' Likewise, in *The Farthest Shore*, the Master Patterner asks Ged, 'What is evil?' and Ged replies, 'a web we men weave' (312).

Recognition, per Clute, requires that 'protagonists begin to understand what has been happening to them [...] They understand, in other words, that they are in a Story; that [...] their lives have the coherence and significance of Story' (Clute 1999c, 'Recognition'). Recognition is not merely a realisation of the problem, then, but an acknowledgement and comprehension of a role in terms of the trajectory of the narrative. Carson, however, has no moment of recognition: she strands her readers in the thinned world, as she is not writing fantasy, and for her the completion of the hero's journey must occur in the realm of politics.

Finding utopia after *Silent Spring?*

A further reason for considering less 'literal' approaches to gaining insight about the future of humankind and its relationship with the non-human world may be emerging: as Carson and Le Guin suggest, the structure of the novum – a physical change that changes society – relies upon notions of prediction and control that have somewhat lost their utopian sheen. Brian Stableford's analysis of 'future histories,' the most 'predictive' subgenre of SF, notes 'a dramatic decline in optimism' across the past quarter century, a symptom of the exigencies of the 'Utopian prospectus, [which,] however tentative, had by then to take problems of ecological sustainability into account' (2007, 136). By the beginning of this millennium 'the great majority of science-fiction images of the future [took] it for granted that the ecocatastrophe was not only under way but already irreversible' (140). The task of imagining a future both plausible and *desirable* is not an easy one these days. Even Suvin, 'having arrived within hailing distance of the end of our species and perhaps of vertebrate life on Earth,' finds that 'the most daring utopia [...] is today [...] not Earthly Paradise but the prevention of Hell on Earth' (1998, 185–186). If, as Frederick Buell puts it, the foregone conclusion of global environmental change seems to shackle the utopian imagination, implying an 'indebtedness to, not a break with, the past,' then literary criticism (with its privileged access to the less-plausible but perhaps more-utopian parts of the human imagination) has a vocation in going beyond the 'literal' and the 'scientific' aspects of SF to uncover the roles played by culture, art and the imagination in creating a better, more liveable, or more cooperative world (2003, 222).

This is precisely what Le Guin's reinterpretation of Carson's story attempts, using the elements of the original to interrogate different notions of utopia which correspond to different ways of 'seeing' the world. As Ged and Arren travel from town to town to discover why True Speech is being forgotten, Arren says:

> It's that way with everything; they don't know the difference [...] They complain about bad times, but they don't know when the bad times began; they say the work's shoddy, but they don't improve it [...] It's as if they had no lines and distinctions and colors clear in their heads. Everything's the same to them; everything's grey (379)

Carson identifies the same problem: in language reminiscent of the villagers in *The Farthest Shore*, who 'go about [...] without looking at the world' (305), she asks, 'Have we fallen into a mesmerized state that makes us accept as inevitable that which is inferior or detrimental, as though having lost the will or the vision to demand that which is good?' (22). Famously, she uses the words of ecologist Paul Shepard, accusing society of

> idealiz[ing] life with only its head out of water, inches above the limits of toleration of the corruption of its own environment...Why should we tolerate a diet of weak poisons, a home in insipid surroundings[? W]ho would want to live in a world which is just not quite fatal? (22)

In this passage, which is one of very few in which Carson criticises the public instead of figuring them as pawns of the chemical companies' manipulations, there lurks the suggestion that the thinning of the world has been, if not brought about, at least enabled by the 'mesmerised' indifference of her readers, who seem to have forgotten the words that describe utopia. Threatened with epistemological homelessness in a thinned and polluted space we cannot understand, how ought we to recover a 'place in the world?' In Carson's terms, what must be done to recover 'will' and 'vision' and 'demand the good?' In other words, where is utopia to be found now?

Le Guin's reduced world

In *Silent Spring*, the chemical industry's 'crusade to create a chemically sterile, insect-free world' can easily be seen as a utopian project of world-transformation: 'Nature has introduced great variety into the landscape, but man has displayed a passion for simplifying it' by killing whichever insects and planting whichever variety of plant seems most convenient for narrowly defined economic use (22, 20). What emerges in Le Guin's narrative is that an evil sorcerer, Cob, has tried to create a place where death is abolished, where people are free of the limits that are imposed upon the rest of the natural world. At first one might read this departure from the language of balance and into the language of immortality as a divergence between Le Guin and Carson, but Le Guin's descriptions encourage a reading that places Cob in particular relation to the discourse of environmental crisis. For example, he states, 'Let all stupid nature go its stupid course, but I am a man, better than nature, above nature' (461). For this reason, the utopia Cob has created is finally located in 'the dry land,' where there is no life, only dust and roaming human souls. Cob's utopia, then, is a 'reduced world,' where nature has been removed to make room for an imagined paradise. Cob's followers 'may return,' after death to 'walk upon the hills of life,' but, we learn, 'their eyes [catch] no light of the moon [and] they cannot stir a blade of grass' (447). Human beings gain eternal life and escape 'the horror of death,' but at the cost of estrangement from the world around them.

However, this idea of simplifying the world is also connected by Le Guin to the idea of cognitive estrangement, perhaps in reference to 'world reduction,' a Marxist SF term she was likely aware of, as Fredric Jameson, one of the theorists most sympathetic to Suvin's utopian framework, uses several of her texts to develop the notion. In 'world reduction,' 'our being-in-the-world is simplified to the extreme[,] abstracted so radically as to vouchsafe, perhaps, some new glimpse as to the ultimate nature of human reality' (1975, 222). Darko Suvin makes a similar comment when he links SF to the pastoral tradition, calling the latter 'an early try in the right direction' because of the exclusion of urban, economic, and national factors from its reality, which 'allows it to isolate, as in the laboratory,' certain 'human motivations' (1979, 9). In other words, despite the premium on 'reality' in radical utopian thought, this reality is of a specialist nature. The environment of a utopia is 'real,' but real in the manner of a petri dish or vacuum chamber.

As we have seen, the route to Cob's 'simplified' utopia comes at a price, and that price is naming: 'Nobody can take his name through. The way is too narrow' (392). In order to reach Cob's utopia, 'you forget the names, you let the forms of things go, you go straight to the reality. [...] A name isn't real, the real, the real forever' (349). By contrasting naming with 'the real forever,' Le Guin suggests that the 'scientific' but abstract reality of cognitive estrangement is only one kind of reality, a Euclidean perspective which 'reduces' the world and thus makes it impossible to imagine a utopia based upon what is actually there.

> 'This is the place,' [Cob] said at last, a kind of smile forming on his lips. 'This is the place you seek. See it? [...] We shall be kings together.' [...] It was wide and hollow, but whether deep or shallow there was no telling. There was nothing in it for the light to fall on, for the eye to see. It was void. Through it was neither light nor dark, neither life nor death. It was nothing. It was a way that led nowhere. (465)

Le Guin has phrased this almost identically in her non-fiction: for her, the Euclidian Utopia is 'nowhere. It is pure structure without content; pure model; goal. [...] Utopia is uninhabitable.' An inhabitable utopia is one which is named, engaged with on its own terms, a centre of the world not abstracted away from the material realm and into the theoretical one. In her utopian science fiction texts, Le Guin makes use of technological advances and scientific 'nova,' which she clearly sees as a way to improve human lives. However, what *The Farthest Shore* suggests is that no amount of technological intervention, necessary as it may be, is sufficient to bring about utopia without an attempt to engage seriously with reimagining a human place in the world. Imagining that utopia is just a novum away might foreclose the perception of possibilities inherent in the world we have.

Rather than propounding world reduction, Le Guin argues that '[t]o reconstruct the world, to rebuild or rationalize it, is to run the risk of losing or destroying what in fact is' (1989, 3). Ged makes the same criticism of Cob, when the latter discovers that he himself no longer has a name: 'You sold the green earth and the sun and stars to save yourself. But you have no self. All that which you sold, that is yourself. You have given everything for nothing' (463). Le Guin describes utopia in her non-fiction as any place one could name: a centre of the world. The human place in the world, which is missing from Carson's narrative, is everywhere: but only if the humans in question are not searching for a 'better world,' dreaming of utopias that are estranged and cognitive rather than here and now.

World multiplication, a utopian alternative

The Farthest Shore suggests that it is not only the earth that needs to be rehabilitated, but the idea of 'the good place' that supports engagement with the world. In times of ecological crisis, the world is thinned not only because of the destruction of the environment but because of the destruction of a sense that human beings have a place in it. Le Guin suggests that yearning for 'another world,' one which is 'reconstructed' or 'rationalised,' is giving everything for nothing. When Arren sees a landscape destroyed by Cob's 'simplification,' an image of ecological destruction, he says it looks 'dead' to him. But Ged chastises him: 'Look at this land; look about you. [...] The hills with the living grass on them, and the streams of water running. [...] In all the world, in all the worlds, in all the immensity of time, there is no other like each of those streams' (449).

> [Then Arren] saw him for the first time whole, as he was. [He] saw the world now with his companion's eyes and saw the living splendor that was revealed about them in the silent, desolate land, as if by a power of enchantment surpassing any other, in every blade of the windbowed grass, every shadow, every stone. (449)

The language of this description unifies the human role – Ged 'whole, as he was' – and the unity of the space he perceives and can name: it goes on to say that Ged saw the landscape as if it were his home, to be seen for the last time. The power of enchantment, which makes a utopia even in environmental crisis, does not inhere in the landscape, which has neither familiarity nor beauty, but in Ged's level of engagement with the process of naming his world, of discovering that 'there is no other like each of those streams.'

Most genre fantasy narratives separate the moment of recognition – in which the hero sees his role in the story, and in the world – and the moment of healing, in which the world is restored. In this passage, Le Guin combines the two. Thus, instead of the hero's journey completing the landscape, the hero's (perspectival) completion of the landscape completes his journey. He is in a place he can name, a centre of the world. The revolutionary utopian definition of 'science fiction,' as a way of reading, defines human society as the dependent variable, and sorts the rest into independent variables (a novum, or a matter of cold fact). Its utopias are 'alternative worlds,' and their 'possibility' is contingent on the belief that we have everything mapped out, that we are in control. In environmental crisis, it no longer seems simple to control, predict, or alter the world to meet human specifications, which is why utopias of the possible seem to break down. Clute writes that science fiction's crucial moment is a 'conceptual breakthrough,' a discovery of something new that remakes the world, while, by contrast, Fantasy's is Recognition: 'an acknowledgement that one has been there all the time' (Clute 1999c, 'Recognition'). What Le Guin adds to the notion is the particularly literary – rather than 'literal' – methodology that 'recognising' utopia requires: to find out the names of things, to discover one inhabits (and always has) a story-shaped world. Utopia is found by finding the world, and (in doing so thoroughly) finding one's role in it. Le Guin has said it well: 'Words hold things. They bear meanings. A novel is a medicine bundle, holding things in a particular, powerful relation to one another and to us' (1996, 153).

Notes

1. *The Farthest Shore* is a member of Le Guin's original Earthsea trilogy, which was followed by three more books after a feminist 'revisioning.' This and all further citations of the Earthsea texts (except where specified otherwise) are from *The Earthsea Quartet*, published by Penguin Books in 1993, which contains the first four books of Le Guin's Earthsea cycle.

2. In this article, I will be using the term SF, a valuable neologism because of its vagueness, which allows it to refer to the complex of texts, modes, and production apparatus associated with 'speculative fiction,' 'science fiction,' 'structural fabulation,' etc., and to sidestep the otherwise entangling debates over murky genre boundaries between a baffling array of designations such as 'science fantasy.'

3. For explicit articulations of this, see Adeline Johns-Putra's (2010) 'Ecocriticism, genre, and climate change,' Trexler's and John's-Putra's (2012) review of 'Climate change in literature and literary criticism,' and much of Ursula Heise's recent work, including a few recent journal articles listed in my works cited.

4. The two-volume *Theory in the Era of Climate Change* series from the Open Humanities Press provides a wealth of excellent engagements with this set of ecocritical discourses. For further discussions of planetary scale, see Ursula Heise's (2008) *Sense of Place and Sense of Planet*, while Trexler and Johns-Putra (2012) provide a useful summary of the temporal and epistemological issues associated with climate change. Timothy Morton's work offers another (and early) example of an attempt to come to terms with the novelty and challenges associated with ecological crisis.

5. In my experience, those who do not cite Suvin cite Fredric Jameson, one of Suvin's most prominent interlocutors within the Marxist Utopian SF school of thought.

6. The diversity of such approaches has been extensively explored and theorised in SF studies. This ranges from Brian Aldiss arguing that SF, in depicting humanity in relation to a changing world, is inherently 'environmental literature,' to more focused approaches such as Eric Otto's (2012) recent *Green Speculations,* about environmentally themed SF texts. A brief but suggestive analysis of the 'missed connections' between ecocritics and SF studies scholars was outlined by Chris Pak (2012) in the Fall 2012 *SFRA Review*.

7. For a brief discussion of the post-war rejection of 'scientifiction' and its uncanny ability to enthusiastically foresee scientific advances (but not their unfortunate social impacts), see Chapters 3, 4, and 12 of *A Companion to Science Fiction*, ed. David Seed (Stableford 2007), or any longer history of SF's origins. Some definitions of 'hard SF' may be more exclusive than Suvin's but the former at least declares itself as a sub-genre, leaving room in the 'soft SF' camp for others.

8. A brief discussion of this in relation to critiques from fantasy author China Mieville and critic Mark Bould appears in William Burling's 'Marxism and SF' chapter, cited in my references.

9. This is complicated by the fact that Suvin defines Utopia as a sub-genre of SF. Here I am arguing that the desire for possible worlds – and Suvin insists there is no alternative world that is not either Eutopian or Dystopian – could easily be defined as a utopian one, regardless of how Suvin classifies the term.

10. It should be noted that Suvin's position on fantasy – and especially Le Guin's fantasy – is complex. In a single essay, he has used the original Earthsea trilogy simultaneously as an example of 'the best the genre has to offer,' and as a means to exemplify why fantasy is nonetheless lacking in the value he finds in science fiction. His view, while more moderate than it was initially, has never relinquished ground on the importance of speculating 'possible worlds.' In the essay cited above, he expressed admiration for Le Guin's feminist 'revisions' to the original Earthsea vision in the later texts *Tehanu* and *Tales from Earthsea*. However, he failed to pardon the original trilogy (the subject of this essay) and persisted in criticising magic and dragons as lacking in the crucial cognitive component he seeks.

11. Many have pointed out that such a Campbellian approach to literature is both restricting and unfashionable. Many others have said so, and then conceded that in the case of the 'very specific subgenre variously referred to as Epic Fantasy, High Fantasy or Heroic Fantasy,' of which Earthsea is a member, such a definition still seems to apply (Duncan 2008, para. 16). I in no way wish to prescribe a definition of fantasy as it is not at all obvious that any literary genre either can or should be constrained by a structural approach. In my reading, however, *The Farthest Shore* not only manifests Clute's structure, but relies on the reader's recognition of these generic conventions to make its impact.

References

Aldiss, B., and D. Wingrove. 1986. *Trillion Year Spree: The History of Science Fiction*. New York: Atheneum.

Attebery, B. 1992. *Strategies of Fantasy*. Bloomington: Indiana University Press.

Attebery, B. 2013. "Structuralism." In *The Cambridge Companion to Fantasy Literature*, edited by E. James, and F. Mendlesohn, 79–90. Cambridge: Cambridge University Press.

Bean, W. 1968. *California: An Interpretive History*. New York: McGraw-Hill.

Buell, F. 2003. *From Apocalypse to Way of Life: Environmental Crisis in the American Century*. London: Routledge.

Buell, L. 1998. "Toxic Discourse." *Critical Inquiry* 24 (3): 639–665.

Burling, W. 2009. "Marxism and SF." In *The Routledge Companion to Science Fiction*, edited by M. Bould, A. M. Butler, A. Roberts, and S. Vint. London: Routledge.

Carson, R. 1962. *Silent Spring*. Crest Reprint. New York: Fawcett Publications.

Clute, J. 1999a. "Landscape." In *The Encyclopedia of Fantasy*, edited by J. Clute, and J. Grant. London: Orbit.

Clute, J. 1999b. "Fantasy." In *The Encyclopedia of Fantasy*, edited by J. Clute, and J. Grant. London: Orbit.

Clute, J. 1999c. "Recognition." In *The Encyclopedia of Fantasy*, edited by J. Clute, and J. Grant. London: Orbit.

Clute, J. 1999d. "Wrongness." In *The Encyclopedia of Fantasy*, edited by J. Clute, and J. Grant. London: Orbit.

Clute, J. 1999e. "Land." In *The Encyclopedia of Fantasy*, edited by J. Clute, and J. Grant. London: Orbit.

Clute, J. 1999f. "Dark Lord." In *The Encyclopedia of Fantasy*, edited by J. Clute, and J. Grant. London: Orbit.

Clute, J., and P. Nicholls, eds. 2012. "Le Guin, Ursula K." *The Encyclopedia of Science Fiction*. Gollancz. Accessed January 3, 2013. http://sf-encyclopedia.com/entry/le_guin_ursula_k.

Cohen, T. 2012. *Theory in the Era of Climate Change: Telemorphosis*, vol. 1. Ann Arbor, MI: Open Humanities Press.

Csicsery-Ronay, I. 2003. "Marxist Theory and Science Fiction." In *The Cambridge Companion to Science Fiction*, edited by E. James, and F. Mendlesohn. Cambridge: Cambridge University Press.

Duncan, H. 2008. "Narrative Grammars." *Notes from the Geek Show*. Accessed March 5, 2013. http://notesfromthegeekshow.blogspot.co.uk/2008/01/narrative-grammars.html.

Elliott, R. C. 1970. *The Shape of Utopia*. Chicago, IL: University of Chicago Press.

Garrard, G. 2012. *Ecocriticism*. London: Routledge.

Gifford, T. 1999. *Pastoral*. London: Routledge.

Heise, U. K. 2008. *Sense of Place and Sense of Planet: The Environmental Imagination of the Global*. New York: Oxford University Press.

Heise, U. K. 2012a. "Reduced Ecologies." *European Journal of English Studies* 16 (2): 99–112.

Heise, U. K. 2012b. "The Invention of Eco-Futures." *Ecozon@: European Journal of Literature, Culture and Environment* 3 (2): 1–10.

Jameson, F. 1975. "World-Reduction in Le Guin: The Emergence of Utopian Narrative." *Science Fiction Studies* 2 (3): 221–230.

Johns-Putra, A. 2010. "Ecocriticism, Genre, and Climate Change: Reading the Utopian Vision of Kim Stanley Robinson's Science in the Capital Trilogy." *English Studies* 91 (7): 744–760.

Killingsworth, M. J., and J. S. Palmer. 1992. *Ecospeak: Rhetoric and Environmental Politics in America*. Carbondale: Southern Illinois University Press.

Krech, S. 1999. *The Ecological Indian: Myth and History*. New York: W.W. Norton.

Kroeber, A. L. 1976. *Handbook of the Indians of California*. New York: Dover.

Le Guin, U. K. 1968. *A Wizard of Earthsea*. Berkeley, CA: Parnassus Press.

Le Guin, U. K. 1979. *The Language of the Night: Essays on Fantasy and Science Fiction*, edited by S. Wood. New York: Putnam.

Le Guin, U. K. 1989. "A Non-Euclidean View of California as a Cold Place to Be." In *Dancing at the Edge of the World: Thoughts on Words, Women, Places*, 80–100. New York: Grove Press.

Le Guin, U. K. 1993. *Earthsea Quartet*. London: Penguin Books.

Le Guin, U. K. 1996. "The Carrier Bag Theory of Fiction." In *The Ecocriticism Reader: Landmarks in Literary Ecology*, edited by C. Glotfelty and H. Fromm, 149–154. Athens: University of Georgia Press.

Murphy, P. D. 2001. "SF and Ecocriticism." In *Beyond Nature Writing: Expanding the Boundaries of Ecocriticism*, edited by K. Armbruster, and K. R. Wallace. Charlottesville: University of Virginia Press.

Otto, E. C. 2012. *Green Speculations: Science Fiction and Transformative Environmentalism*. Columbus: Ohio State University Press.

Pak, C. 2012. "Composting Culture: Literature, Nature, Popular Culture, Science University of Worcester, 5–7 September 2012." *SFRA Review* 302: 4–5.

Philmus, R. M. 2005. *Visions and Re-visions: (Re)constructing Science Fiction*. Liverpool: Liverpool University Press.

Rochelle, W. 2001. *Communities of the Heart: The Rhetoric of Myth in the Fiction of Ursula K. Le Guin*. Liverpool: Liverpool University Press.

Sawyer, A. 2006. "Ursula Le Guin and the Pastoral Mode." *Extrapolation* 47 (3): 396–416.

Stableford, B. 2007. "Science Fiction and Ecology." In *A Companion to Science Fiction*, edited by D. Seed. Oxford: Blackwell.

Sussman, H. 2012. *Theory in the Era of Climate Change: Impasses of the Post-Global*, vol. 2. Ann Arbor, MI: Open Humanities Press.

Suvin, D. 1979. *Metamorphoses of Science Fiction: On the Poetics and History of a Literary Genre*. London: Yale University Press.

Suvin, D. 1998. "Utopianism from Orientation to Agency: What Are We Intellectuals Under Post-Fordism to Do?" *Utopian Studies* 9 (2): 162–190.

Suvin, D. 2006. "On U. K. Le Guin's 'Second Earthsea Trilogy' and Its Cognitions: A Commentary." *Extrapolation* 47 (3): 488–504.

Trexler, A., and A. Johns-Putra. 2012. "Climate Change in Literature and Literary Criticism." *Wiley Interdisciplinary Reviews: Climate Change* 2 (2): 185–200.

The Biologisation of Ecofeminism? On Science and Power in Marge Piercy's *Woman on the Edge of Time*

Martin Delveaux

Utopia, the *Penguin Dictionary of Literary Terms and Literary Theory* (1999: 957–960) informs us, is a place where all is well. Coming from the Greek *eu*, meaning 'good', and topos, meaning 'place', Utopia evokes associations of a paradisiacal society where people enjoy equal rights and live in harmony with their environment. This is particularly true of the Ecofeminist Utopia, a utopian novel which is influenced by feminist and environmentalist ideas and practices.

Explaining the origins of Ecofeminism, the ecocritic Richard Kerridge (1998: 6) notes that in the 1980s, sparked off by recurring ecological disasters, such as the Chernobyl disaster in the former Soviet Union, there emerged an important body of thought in Feminism which argued that the beliefs and institutions which oppress women were largely those which cause environmental damage and that Feminism and Ecology could make common cause under the heading 'Ecofeminism'.

First introduced in 1974 by the French feminist Francoise d'Eaubonne, who was writing about women's role in a proposed ecological revolution (cf. Graham, 1998: 115), the term Ecofeminism can be considered an umbrella term which captures a variety of multicultural perspectives on the connections between those humans in subordinate positions, particularly women, and the domination of non-human nature (Warren, 1994: 1). Ecofeminist Utopias, accordingly, depict a society free of misogyny and environmental destruction, and suggest a life-style 'in the balance'.

Marge Piercy's *Woman on the Edge of Time* can be read as an Ecofeminist Utopia because Piercy's idealised vision is based on a future society without misogyny and environmental exploitation. Praised for exemplifying the "most influential of ecofeminism's roots" (McGuire/McGuire, 1998: 196). Piercy's novel has reached the status of "a contemporary classic" (Booker, 1994: 337). Published in 1976, *Woman on the Edge of Time* tells the story of Connie Ramos, a 37-year old Mexican-American, who lives at the bottom of American society in New York in 1976. After being signed in to a mental hospital under conspirational circumstances, Piercy's anti-hero becomes an experimental subject for the male doctors. The only successful way Connie can escape from the ward is through her mental abilities, as she is able to get into contact with Luciente from Mattapoisett, a utopian future society which lives without misogyny and environmental exploitation.

It is the description of this ecofeminist counter-society of Mattapoisett that establishes Piercy's importance in the canon of ecofeminist utopian literature. The feminist scholar Kerstin Shands (1994: 72) rightly points out that in this alternative future it is the ideas of feminism, at a height when *Woman on the Edge of Time* was written, that are realised, and that ecological concerns are paramount. Indeed, Piercy offers a whole series of environmentally friendly ideas and practices in Mattapoisett as a model for emulation: wind mills are used to produce energy and the inhabitants do not waste anything and make use of natural, degradable building materials, such as stone and wood. Work has been freed from the economic pressures of capitalism and is now associated with art (which is strongly reminiscent of William Morris's utopia *News from Nowhere*, published in 1891), and recycling has achieved a status of prime importance in Mattapoisett. Children are brought up in 'scavenger works crews', and the level of importance the people of Mattapoisett ascribe to the natural environment is also mirrored on the political level, where every township has its own Earth Advocate, who speaks for the rights of the environment. The Earth Advocate (whose hair colour is, significantly, green) and the environmentally friendly attitude of the people in Mattapoisett have contributed to a landscape that is suffused with gardens and other green areas and is evocative of a bucolic, Arcadian landscape:

> Most buildings were small and randomly scattered among trees and shrubbery and gardens, put together of scavenged old wood, old bricks and stones and cement blocks. Many were wildly decorated and overgrown with vines. [. . .] In the distance beyond a blue dome cows were grazing, ordinary black and white and brown and white cows chewing ordinary grass past a stone fence. Intensive plots of vegetables began between the huts and stretched into the distance. (pp. 68–69).

The environmentally friendly attitude of the people in Mattapoisett is accompanied by a policy of gender-equality. Sex-related work, stigmatisation of homosexuality and legal bonds of marriage do not exist: in Mattapoisett people have 'hand friends' and 'pillow friends' they stay and live together with, but without any legal obligation. In addition, since all factories are completely automated, the traditional division of labour has been abolished. Gender equality is also reflected in terms of language: by using the word 'per', the abbreviated form of 'person', the people from Mattapoisett have replaced the personal pronouns 'he' and 'she' and the possessive pronouns 'his' and 'her'. Against this background, it becomes clear why Piercy's novel is so attractive to ecofeminists, and why Lucy Sargisson (1996: 165), in her book *Contemporary Feminist Utopianism*, has called *Woman on the Edge of Time* a veritable "paradise on earth".

It is important to realise, however, that Piercy's establishment of a society free of misogyny and environmental destruction is dependent on a strongly technophilic culture. To gain natural energy, for example, the roofs of the houses are fitted with rainwater-holding and solar energy facilities. Likewise, in transport, solar airboats, so-called dippers, are used which ride on a cushion of air and function as a communal means of transport. The extent to which Luciente's society has been technologised is also indicated by the fact that they consider their *kenner*, a sort of mobile phone that they carry on their wrist, as an integral part of their body. Without their kenner, they feel naked (p. 327). In Mattapoisett, the use of technology has such a strong impact on the formation of their cultural identity that the boundaries between human and technology become blurred. Importantly, this is also emphasised in the context of equality of the sexes. In a particularly bold subversion of conventional sex contrasts, the men in Mattapoisett are biologically able to nurse babies. The narrator explains about one of the male characters:

> He had breasts. Not large ones. Small breasts, like a flat-chested woman temporarily swollen with milk. Then with his red beard, his face of a sunburnt forty-five-year-old man, stern visaged, long-nosed, thin-lipped, he began to nurse. The baby stopped wailing and begun to suck greedily. (p. 134)

Despite the fact that people in Mattapoisett have one primary sexual organ and are not hermaphrodites, the phenomenon of male breast-feeding demonstrates that Piercy's novel is informed by a heightened level of cross-gendered social and biological roles. This 'mild' form of androgyny shows that Piercy envisages a less oppositional culture between the sexes.

Here, Piercy clearly takes a political-ecofeminist stance. As Maria Mies and Vandana Shiva have observed (cf. 1993: 18), there is an important divide within Ecofeminism between spiritual (or 'cultural') ecofeminists and political ecofeminists. While the approach of Spiritual Ecofeminism is to "stress the links, historical, biological, and experiential, between women and nature, and see their joint oppression as the consequence of male domination" (Plumwood, 1993: 10). Political Ecofeminism does "not see women's difference as either [sic] biologically determined in their relationship to nature or to one another." (10). Whereas this latter perspective sees nature as a political rather than a natural category, and proposes the construction of a less oppositional culture, Spiritual Ecofeminism hails the creation of an alternative women's culture as a solution to the bi-dimensional oppression. Instead of abandoning this dualistic tradition, Spiritual Ecofeminism seeks to revalue what patriarchy has devalued, trying to turn a negative into a positive.

In trying to overcome problems of sexism and environmental exploitation, Piercy clearly adopts a political-ecofeminist view. Rather than trying to revalue what patriarchy has devalued, the people of Mattapoisett have developed a less oppositional culture between the sexes in which people's subjectivities are not seen as biologically determined. However, this less oppositional culture has been achieved through a heightened technologisation and control of the reproductive processes. In Mattapoisett, reproduction is ensured by "Mother, the machine" (p. 102), a so-called 'brooder', which stores all the genetic material of the future embryos. With this relocation of reproduction, the society of Mattapoisett largely corresponds, as Lucy Sargisson has observed (1996: 165), to the claims postulated by the feminist scholar Shulamith Firestone. One of Firestone's main arguments is that misogyny is rooted in reproduction, since woman's ability to give birth makes her dependent on man. In *The Dialectic of Sex*, she claims that

> women, biologically distinguished from men, are culturally distinguished from 'human'. Nature produced the fundamental inequality—half the human race must bear and rear the children of all of them—which was later consolidated, institutionalized, in the interests of men. (1971: 232)

Since women, according to Firestone, can only be completely independent if they are freed from their biological role as mothers, she demands the freeing of women from the tyranny of their reproductive biology by every means available, and the diffusion of the child-bearing role to the society as a whole, men as well as women (1971: 233). In a striking parallel, in *Woman on the Edge of Time* one of the basic preconditions for establishing a society based on gender equality is to abolish reproductive difference; sexual difference is reduced to such an extent that it borders on androgyny.

The reasons behind the faith in the technologisation of reproduction in Piercy's novel must be seen in the historical context of *Woman on the Edge of Time*. As the cultural historian Angelique Richardson (cf. 2000: 40–41) has convincingly argued, historically, biology has not been a friend to women. In the second half of the nineteenth century, for example, the anthropologist Paul Broca made much of women's smaller brains in *The Anthropological Review* (1868: 635–652). Three years later, Charles Darwin declared in *The Descent of Man* that "man is more courageous, pugnacious and energetic than women, and has a more inventive genius", whereas woman has stopped "intermediate between the child and the man" (1981: 616, 327). Similarly, Herbert Spencer argued that women had failed to evolve as completely as men on account of their reproductive function (1873: 30–38). During the second half of the twentieth century, however, biology—at least to some extent—seemed to hold the answers to woman's problems. As Cathleen and Colleen McGuire (1998: 197) have observed, in the 1970s, at a time when the birth control pill helped the women's liberation movement take off the ground, many women looked to science for further solutions to their problems. If bio-technology were wrested from the oppressive hands of patriarchy, so the dogma held, then the fruits of science could liberate women and in turn society, such as in the androgynous society of Mattapoisett.

It has been argued, however, that androgyny is not intrinsically and necessarily synonymous with equality. Luce Irigaray (1993: 12), for example, one of the leading 'difference feminists', has argued in her book *Je, Tous, Nous: Toward a Culture of Difference* that "[w]omen's exploitation is based upon sexual difference; its solution will come only through sexual difference . . . To wish to get rid of sexual difference is to call for a genocide more radical than any form of destruction there has ever been in History." Similarly, I would argue that in Piercy's novel the biological equality of the sexes has not led to egalitarian power relations, which emphasises the importance of Edward Said's claim that "[i]deas, cultures, and histories cannot seriously be understood or studied without their force, or more precisely their configurations of power, also being studied" (1978: 5). In Piercy's fictional society, the gain of women's ecofeminist ideals, of independence and equal gender relationships, has been, I would argue, at the expense of other power relationships, particularly the relationship between the individual and society. Reminding the reader of Huxley's *Brave New World*, the narrator gives a dystopian description of the 'bottle-babies':

> A door slid aside, revealing seven human babies joggling slowly upside down, each in a sac of its own inside a larger fluid receptacle. Connie gaped, her stomach also turning slowly upside down. All in a sluggish row, babies bobbed. Mother the machine. Like fish in the aquarium at Coney Island. (p. 102)

Especially the fact that in Mattapoisett 'mother', in the biological sense, is equated with 'machine' must be seen as problematic in terms of egalitarian power-relations, as this form of power-sharing is concomitant with a limitation of individuality and personal wishes. Through the mechanising of reproduction, exact birth- and population control has become possible, and one of the basic rules in Mattapoisett is "One in, one out" (p. 162). Furthermore, the future characteristics of a person to be born can be determined *a priori* by feeding the genetic information into the machine. Luciente foresightedly explains that they are all a "mixed bag of genes" (p. 100). Although it is claimed that nurture counts more heavily than genetics (p.324), the reader learns that this rule only applies "once you've weeded out the negative genes" (p.324), clearly revealing the approval of eugenics in Mattapoisett. In spite of democratic structures in terms of taking decisions—the representatives of every township are not elected, but chosen by lot (p. 151)—the possibility of moulding a human being *a la carte* holds many dangers of abuse of power. As in Huxley's *Brave New World*, the appropriation and manipulation of this machine would entail an enormous shift of power to those in charge of the machine. The extent to which the machine is already used to manipulate the natural dispersion of ethnic groups is shown by the fact that the people of Mattapoisett, aiming to eliminate racism, are breeding a higher proportion of darker-skinned people. However, Luciente admits that they don't want a 'melting pot' and, therefore, decided to hold on to separate cultural identities (p. 104). Thus, it becomes obvious that reproduction is an effective tool for social control in Mattapoisett.

Although it seems that racism has ceased to exist in Mattapoisett, it needs to be pointed out that the elimination of racism has been achieved through a heightened sense of racial separatism. However, as Angelique Richardson (2000: 38) has argued, a progressive, truly radical feminism, remembering that we are interdependent inhabitants of an increasingly small planet, would favour integration, not separatism.

Since the power of human reproduction, the guarantee of the survival of humankind, has been shifted to a machine, humanity is completely dependent on the proper functioning of the machine. If, however, the machine stops, as in E.M. Forster's dystopian short story *The Machine Stops* (1909), then humanity is doomed to perish: one realises that gender equality in Mattapoisett is based on a very fragile foundation, where reproduction has little in common with the *conditio humana*. Instead of giving men *and* women the chance to give birth, Piercy completely de-humanises reproduction. Although, admittedly, people do lead egalitarian relations in terms of gender, I would question whether the equality between the sexes might not be established without such a drastic intervention in the human genetic stock which sacrifices human independence.

Undoubtedly, Piercy aims at eliminating the cultural, frequently deterministic significance of biology. As one of the characters in Mattapoisett explains, "we broke the bond between genes and culture" (p. 104). It seems ironic, however, that the elimination of the cultural significance of biology is not achieved through the means of culture but biology— through an increased human interference in the biological processes of the body (e.g. the use of eugenics and the 'brooder' for biological modelling). Consequently, since Mattapoisett has overcome misogyny and environmental exploitation by *biologically* breaking the supposedly oppressive nature of heterosexual relations, Piercy's novel fails to show how to overcome the social structures of oppression. In *Woman on the Edge of Time*, I would argue, Ecofeminism has been biologized to such an extent that its social meaning becomes eclipsed.

Although with the help of the brooder future characteristics of the person to be born can be determined, there is still a debate in Mattapoisett over the issue of selective breeding. As Luciente explains (p. 226), the general debate has polarised between the *Shapers*, who want to breed for selected traits, and the *Mixers*, who watch for birth defects and fix the proper gene balance (p. 262). The mere fact that people discuss whether to support the *Mixers* or the *Shapers* indicates the broad acceptance of genetics which, in turn, reveals the anthropocentric world picture of the society. Instead of minimising the human impact on nature, the people of Mattapoisett readily disturb the balance of natural ecosystems by breeding many varieties of vegetables resistant to drought (p. 210). Disturbingly, a similar kind of the 'weeding out of negative genes' in terms of genetics also surfaces in the realm of human relationships. Although Luciente asserts that cooperation is a major feature of her society, she cannot conceal the fact that they still cling to a competitive and selective modus vivendi. As already insinuated by Luciente's views that always some competing goes on (p. 174) and that they have to struggle to exist (p. 197), the selectivity becomes most important in the context of the initiation ritual, where the child is left in the woods for one week where it must learn how to survive. Asked by Connie what happens if the child is bitten by a snake or gets appendicitis, Luciente coldly replies "We take the risk [. . .] You're right, accidents happen" (p. 116). This practice. I would argue, is strongly reminiscent of a Spencerian survival-of-the-fittest attitude that has little in common with what one would expect in a paradise.

Against this background, the presence of the so-called 'drifters' for whom Mattapoisett does not constitute a paradise does not come as a surprise. The drifters do not subscribe to the uniform ideals proposed by the society and, in an atmosphere of cultural phobia, are victimised as scapegoats. Luciente makes very clear what will happen to those who exercise a form of power against the society by not showing up at the township meetings. In that case, as Luciente relates, "Friends might suggest you take a retreat [. . .] If too many in a village cut off, the neighbouring villages will send for a team of investigators" (p. 154). A similar form of social ostracism happens when people do not conform to the work ethos of Mattapoisett, as such people are asked to leave (p. 101). The reader realises that non-conformity in Mattapoisett results in exclusion. Similarly, those who resort to violence are also excluded and not given much space for improvement. In an open stigmatisation, the offenders are marked with a tattoo on the back of their hands to indicate the danger they pose to the community (p.272). Urged to sit apart in guest houses, these offenders are clearly defined as the 'downs' of the society.

Thus, the people of Mattapoisett correspond to what the cultural historian René Girard (cf. 1986) has termed the 'Generative Scapegoat Mechanism', which emphasises the inherent need to scapegoat in order to maintain prevalent power constellations. In his book *The Scapegoat*, Girard claims that human societies are founded on mechanisms of sacrifice, which provide a community with its sense of collective identity and preserve its cultural values. Usually concealed from human consciousness, sacrifice is initiated by scapegoating and stigmatising the supposedly 'other', the 'different', in an attempt to prevent a 'mimetic crisis', a breakdown of all distinctions of representation. Margaret Atwood (1982: 275) underlines that the fear of this mimetic crisis is not only a frequent structural element within utopian societies, but is also liable to affect the reader, since all utopias suffer from the readers' secret conviction that a perfect world would be dull. Such readers, however, need not worry about *Woman on the Edge of Time*, for Piercy's ecofeminist utopia is far from being perfect.

The accumulation of social exclusions culminates if drifters relapse into crime: then the society of Mattapoisett conveniently executes them in an exercise of power over power which seriously undermines the status of Mattapoisett as a paradise. Luciente explains: "Second time someone uses violence, we give up. We don't want to watch each other or to imprison each other. We aren't willing to live with people who choose to use violence. We execute them" (p. 209). The harsh treatment of those who do not want to serve the ecofeminist ideals brings to the fore that equality is not always a positive quality if it is achieved by suppressing or even excluding individuality and imposing an oppressive conformity which makes all differences disappear. At the same time, the phenomenon of the 'drifters' clearly proves that the biological eradication of difference is no blueprint for happiness and social equality. On the contrary, the disappearance of difference causes, in an illustration of Girardian theory, a mimetic crisis in Connie.

Connie realises that the idyll of the place and the alleged perfection can also be, from another point of view, dystopian:

> "How can some kid who isn't related to you be your child?" [Connie] broke free and twisted away in irritation. The pastoral clutter of the place began to infuriate her, the gardens every-place, the flowers, the damned sprightly-looking chickens underfoot. (p. 105)

Like the drifters in Mattapoisett, Connie is one of the 'downs' in New York, and Shands (1994: 66) rightly points out that Connie's story is the story of victimisation. Although she mostly sees Mattapoisett as utopia, Connie is aware of the oppressive conformity and the increased mechanisation of biology. She is disgusted by the 'bottle babies' to such an extent that she starts hating them, making them the scapegoat for her own situation:

She was sitting against the wall on the porch, tears trickling from her eyes. She hated them, the bland bottleborn monsters of the future, born without pain, multicoloured like a litter of puppies without the stigmata of race and sex. (p. 106)

As becomes obvious, equality can also be concomitant with the imposition of an oppressive uniformity that leaves little space for individual wishes and desires. In fact, the abolition of the boundaries between the individual and society, accompanied by sanctioning non-conformity and restricting individuality, is reminiscent of many classic dystopias. Therefore, I would suggest that the Ecofeminist Utopia is not always a purely positive vision of an emancipated and ecologically sustainable world, but can also rely on the notion of exclusion, which is the culmination of unequal power relations. It must be regarded a myth that Ecofeminist Utopias are unanimously based on collective equality and have an egalitarian approach to community, without acknowledging at the same time that this 'equality' goes hand-in-hand with the imposition of an oppressive totalising uniformity.

In her recent book *Has Feminism Changed Science?*, the historian of science Londa Schiebinger has argued that it is time to move away from conceptions of feminist science as empathetic, nondominating, environmentalist, or 'people-friendly'. Instead, as she goes on to argue, it is time to turn "to tools of analysis by which scientific research can be developed as well as critiqued along feminist lines" (1999: 8). It seems to me that *Woman on the Edge of Time* is a good case in point for this appeal since Piercy's ecofeminist novel clearly lacks these specific tools Schiebinger calls for. Therefore, I would argue that Ecofeminism, no less than any other ideological current, is not immune to abusing power and becoming out of touch with its initial aims. Under the cover of the prefix 'eco', which currently seems to enjoy an ethically impeccable status in the liberal public imagination of the Western World, Ecofeminism too readily sells itself as a product with the label of equality and engenders a kind of critical myopia. For instance, in her book *The Good-Natured Feminist: Ecofeminism and the Quest for Democracy*, Catriona Shandilands has equated Ecofeminism with a 'democratic political vision' (1999: xvii). However, Ecofeminism can also ramify into oppressive, intolerant, and racially discriminative forces that disrupt its original basis. It is important, therefore, to be wary of socio-biological 'blueprints' which claim to provide the scientific tools for a 'just' social organisation, and to realise that Ecofeminism is not inevitably freer from the misuse of science and power than any other ideologically driven social movement, but sometimes masquerades as a paradigm of democratic social transformation.

References

Atwood, Margaret (1982), "Marge Piercy: *Woman on the Edge of Time*, Living in the Open", in *Second Words. Selected Critical Prose*. Toronto: Anansi.

Booker, M. Keith (1994), "Woman on the Edge of a Genre: The Feminist Dystopias of Marge Piercy", in *Science Fiction Studies* 21, 3.

Broca, Paul (1868), "On Anthropology", in *Anthropological Review*.

Cuddon, J.A. (ed.), (1999), *The Penguin Dictionary of Literary Terms and Literary Theory*. Fourth Edition. London: Penguin Books.

Darwin, Charles (1981), *The Descent of Man, and the Selection in Relation to Sex* [1871]. Vol. 1. Chichester: Princeton University Press.

Firestone, Shulamith (1971), *The Dialectic of Sex: A Case for Feminist Revolution*. St. Albans: Paladin.

Forster, E.M. (1950), "The Machine Stops" [1909], in *Collected Short Stories*. London: Reader's Union.

Girard, René (1986), *The Scapegoat*. Trans. Yvonne Freccero. London: Athlone.

Graham, Amanda (1998), "*Herland*: Definitive Ecofeminist Fiction?", in Val Gough and Jill Rudd (eds.), *A Very Different Story: Studies on the Fiction of Charlotte Perkins Gilman*. Liverpool: Liverpool University Press.

Irigaray, Luce (1993), *Je, Tous, Nous: Towards a Culture of Difference*. Trans. Alison Martin. London: Routledge.

Kerridge, Richard, and Sammels, Neil (eds.), (1998), *Writing the Environment – Ecocriticism and Literature*. London and New York: Zed Books.

McGuire, Cathleen and McGuire, Colleen (1998), "Grass-Roots Ecofeminism: Activating Utopia", in Greta Gaard and Patrick D. Murphy (eds.), *Ecofeminist Literary Criticism*. Urbana and Chicago: University of Illinois Press.

Mies, Maria, and Shiva, Vandana (1993), *Ecofeminism*. London and New Jersey: Zed Books.

Piercy, Marge (1979), *Woman on the Edge of Time* [1976], London: The Woman's Press.

Plumwood, Val (1993), *Feminism and the Mastery of Nature*. London: Routledge.

Richardson, Angelique (2000), "Biology and Feminism", in *Critical Quarterly* 42, 3.

Said, Edward (1978), *Orientalism*. London: Routledge and Kegan Paul.

Sargisson, Lucy (1996), *Contemporary Feminist Utopianism*. London and New York: Routledge.

Schiebinger, Londa (1999), *Has Feminism Changed Science?* Cambridge, MA: Harvard University Press.

Shandilands, Catriona (1999), *The Good-Natured Feminist: Ecofeminism and the Quest for Democracy*. Minneapolis and London: University of Minnesota Press.

Shands, Kerstin (1994), *The Repair of the World: The Novels of Marge Piercy*. London & Westport: Greenwood Press.

Spencer, Herbert (1873), "Psychology of the Sexes", in *Popular Science Monthly* 4.

Warren, Karen J. (1994), *Ecological Feminism*. London: Routledge.

Afterword: the utopian dreaming of modernity and its ecological cost

Geoff Berry

Phoenix Institute of Australia, Melbourne, Australia

If utopia has often been criticized as an unrealistic and unsustainable fantasy, then the most prominent example of this imaginary today must be the vision provided by the endless growth and profit model of global capital. This utopian dreaming of perpetual abundance links technological progress with faith in human culture's ability to master the rest of nature. Fredric Jameson followed Ernst Bloch's recognition of this deep yearning for a land of milk and honey and I likewise return to the *Principle of Hope*, this time to trace the way Bloch treats light as a mythic symbol of abundance. Citing the famous NASA photomontage of the planet at night, this article shows how modern urban life relies on the unsustainable combustion of fossil fuels in an effort to banish the night from earthly life. It also cites a selection of advertisements that directly link light with magical yet ubiquitous powers of overcoming, and which ignore or conceal the dire ecological consequences of this Promethean dreaming. The article concludes with reference to the comparatively realistic visions of critical ecotopias and ecodystopias that remind us of our collective hopes for a better future in terms both social and ecological.

Utopia has been criticised as a place unrealistic and unsustainable by conservative commentators determined to deny the imagination of the visionary on behalf of a seemingly more sober estimation of the real and the possible. In the light of the ecological crisis, however, it is the dominant paradigm of production and consumption itself that looks like the manifestation of an impossible dream. The fantasy of the transnational capitalist utopia is generally fudged in the haze of white noise along with which it is extended and marketed; yet it is as startlingly simple – and wholeheartedly as mythic – as any great vision of clarity, hope and purpose. The dominant paradigm represents an unceasing dream of eternal youth in a city of unending plenty. The spectacular successes of industrialised modernity bear this front along like the crest of an ever-breaking wave, while the foundation upon which its materialism is built – the empirical verifiability of stuff – helps to cement an aura of reasoned argument. That this success comes at a high ecological cost, we all know – so why is it so hard to arrest the malignant aspects of postmodern development while maintaining the benign? Exactly because of the combination of material success and fanciful ideal, the satisfied bodily and social desires and the mythic realm of wish fulfilment that accompanies them; in short, because the pleasure principle and its unboundedly selfish craving precedes the reality principle and continually strives to spring past it. But while a psychoanalytic interpretation of the dominant paradigm's utopian dreaming would doubtless yield insightful results, it is Ernst Bloch and his sprawling epic *The Principle of Hope* (1995 [1959]), that most saliently presages this recognition: that the really unsustainable utopian dreaming of today is expressed in the growth and profit model of capitalistic production and consumption.

Bloch identified the West's utopian longing as the wish to overcome death and fulfil every desire by bathing endlessly in a fountain of youth (456–465). Like Fredric Jameson, I follow Bloch's identification of the modern individual's materialistic longing as a new version of the ancient dream of unending plenty, which has been held at least since the inception of agricultural settlement civilisation. Yet while this dream has a well-attested history, Bloch does not diagnose it as necessarily nostalgic. Our wish-fulfilment does not only tend towards a lost golden age of unlimited abundance (either womb- or paradisial garden-like), but also looks out across the horizon. Following the utopian impetus to its locus within human drives, he sees the destination of our collective dreaming conveyed in daydreams and advertising imagery, religious tracts and literary imaginings, as a post-modern as well as timeless image of hope that is both promising and potentially dis-astrous. In the early twenty-first century, we'd have to admit that the companion genre of critical dystopia now offers a similarly realistic appraisal of 'the principle of hope', suggesting grim warnings of the dark future ahead if we do not wake up to dangerous tendencies in the present. After thinking through the way Bloch's ideas might be employed today, I will conclude with a look at the way critical ecotopian and ecodysto-pian imaginaries continue to offer hope for improved relations between humans and the rest of nature.

Light and utopian dreaming; or, Blochian hope at home

The famous NASA photomontage of the planet at night reveals the extent of twenty-first century humanity's transnational urban centres, brilliantly lit against the backdrop of the earth. This image inspires me with beauty and terror, as it gives a stark insight into the extent to which we are lighting up the planet in order to better enjoy our human habitations, burning fossil fuels like there is no tomorrow and turning black gold into transcendent light. This image is the most evocative signal to the environmentalist movement since the first photograph of the planet from space, also an iconic NASA image, which was often subtitled 'the only one we've got' (Berry 2012). Here is the

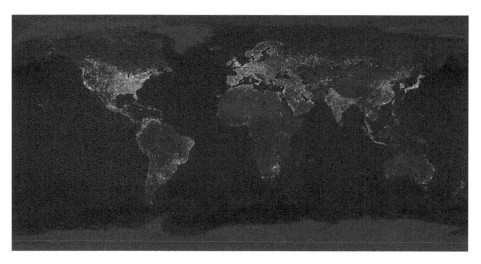

NASA photomontage of the planet at night, showing urban centres brightly lit, 2003 (image courtesy NASA).

planet as co-created, evidence of enormous inventive power but also dangerous hubris, a set of cultural marks that reveals the abstract nature of modern technologies, which act upon the planet as if from above yet are capable of destroying so much in the same dream. This power is paradoxical in its attempt to manifest transcendence even as it acts in denial of our dependence upon the earth (Berry 2013). But it also reveals something much more humble, a tradition we continue to perpetuate regardless of our sociopolitical affiliations – that of the human proclivity to leave our mark as story.

The story of progressively developing technoscience can be seen in the NASA photomontage because it offers, with the clarity of distanced perspective, insight into exactly how the raw material of the earth can be transformed by human culture into a habitation filled with luminescence. As perspective is removed from the immediacy of embodied experience, darkness is transformed into light; yet this represents a paradoxical victory because we are no longer truly at home to enjoy it. Floating free in space, the body becomes just like the earth – raw material used to fuel the hopeless hope of this double-edged quest for transcendence. This dream combines paradoxical twin impetuses of twentieth century philosophies, which operate towards re-embodiment (as in phenomen-ology) but also work to extend the already disembodied scientific (or military industrial) gaze. The planet at night reveals the extent to which we seek mastery over the darkness of raw material, by flooding it with artificial electric light. Bloch is a remarkably prescient commentator of this quest and has even been described (by his translator Neville Plaice) as 'a poet of light' who recognised 'premonitory glimmerings and extended after-glow-ings, shining ahead or continuing to bathe history in their unextinguished light' (Bloch 1995, xxxi). The utopian quality of light works in both physical and metaphorical terms and, most importantly for Bloch, the hope incandescent in the form and image of light unites individual and social spheres. As Moylan notes, Bloch's 'utopian capacity achieves concrete expression only when it is "set on its feet and connected to the Real-Possible" conditions in a particular historical situation' such that the utopian function is rooted in the world and 'is transcendent without transcendence' (1997, 97–98).

Our primal need to live 'in the light', in an embodied sense, is driven by the limits we experience as part of physical life – not just the finality of death but the Kantian categorical boundaries to what we can know, which also lead back towards what Bloch called the 'Darkness of the Lived Moment'. According to this notion, we are never completely present in the here and now, but chase an urge or hope that we can experience complete presence. We can be inspired by a memory from childhood of the feeling of complete immersion in the body and the moment, or from some other analogous sensation of flow. For Bloch, the 'darkness of the lived moment' drives us towards the authentic 'homeland' of our utopian dreaming, which he defines as the 'unpossessed nature of self-location and being-here finally mediated, illuminated and fulfilled' by 'human beings themselves and their environment' (1995, 16). This realization of lack can also, unfortunately, be easily incorporated into the kind of 'quick, thoughtless enjoyment' peddled by the machine of capital (293). This works especially well in denying the ever-present reality of death, which 'must not be remembered, cheap images push it out of mind … [and] false youth is painted on. … The wish is simply to hear and to see nothing of it, even when the end is here. Thus fear at least shrinks, becomes flat, like so much else' (1104–1105).

For Bloch, the 'Not-Yet' realised utopian prospect is both an inner glow and a necessarily collective concern, a signal that draws us towards a *dialectical-materialistically comprehended hope* that must be grasped thoroughly if we are to loosen the bonds of oppressive regimes of fascism or capital (9; italics always in original). Bloch looks ahead, railing against Plato's nostalgic *anamnesis*, which he sees as a conservative recollection

The utopian home: electricity, symbolised (as it so often is) in light (complete with magical fairy). AGL, Australia, 2004.

offering nothing fundamentally new, preferring to place before us his *anagnorisis*, a recognition wherein 'memory traces are reactivated' but with both similarity and dissimilarity 'because of all the intervening novelty' (Geoghegan 1997, 21–23). While the novelty so beloved of capitalist production, with its insistence upon new models, fads and planned obsolescence, is far from the innovation sought by Bloch, it does continue a tendency he had noticed in the 1930s – the investment of human ingenuity in improving physical comforts: 'The whole of life is thus surrounded by a belt of artificial creations which have never existed before. The human house is vastly extended by them, it becomes more and more comfortable and adventurous' (1995, 627).

In this advertisement by Australian power company AGL, the suburban home is imagined as a locus of modern wish-fulfilment. The Blochian idea of the home as both a place of safety and adventure has since expanded exponentially, so that the home is not only illumined with electricity converted elsewhere from fossil fuel to clean, clear light, but also invites the entertainment and propaganda portals of television and, in the laser age of digital communications, virtually immediate transmission from anywhere in the world. The living room couch now becomes a winged armchair, bringing all the dreams of the world together into the home. In this respect it digitally extends psychic space as it collapses the tyranny of distance, working like a hybrid between the flying carpet and the ship of the Argonauts. As Bloch stated, 'every other utopian intention is … indebted to that of geographical discoveries; for each of them has at the centre of its positive hopes the topos: land of gold, land of happiness' (750). The 'darkness of the lived moment' would be expunged in the Blochian moment of Here and Now as Being-For-Itself, the 'All' no longer 'hidden from itself' in a materialization of what would historically have been considered a kind of mystical religious fulfilment (1298). This possibility, like virtually all others spiritual, physical, emotional and mental, has been wholeheartedly co-opted by the dominant paradigm of production and consumption. Indeed, Bloch updated Marx's notion of the commodity fetish to take into account the way capitalism traded on the alienation of urbanised humanity from the rest of nature. Put another way, he saw the way that desires to reproduce a culturally imagined heaven on earth have replaced our traditional intimacy and sympathies with the more-than-human world.

Dominant strands of western culture have continuously attempted to materialise visions of paradise in cites of light. The New Jerusalem of Revelations requires no sun or moon, perpetually lit as it is by the presence of God, a new heaven on earth that replaces the corrupted version. Hans Blumenberg pointed out that traditional myths and metaphors of light commonly supposed a dualism that is overcome in effulgent transcendence; that light battled darkness but then offered itself in complete victory, above and beyond opposition (1993, 32). Similarly Northrop Frye noted that the Biblical narrative form of this model operates according to the comedic U-shape pattern whereby grace is had, lost, and finally regained (2004, 22–23). The mythic New Jerusalem overwhelms resistance with abundance:

> And there will be no more night; they need no light of lamp or sun; for the Lord God will be their light; and they will reign for ever and ever. (Revelation 22:5)

'Enlightenment' philosophies suggested that scientific understandings could replace traditional superstitions, which would in turn help us understand the earth and ourselves in objective terms – a new, humanistic dispensation for a secular people beyond archaic religious beliefs but not necessarily more 'at home' with the rest of life on earth. Ernest Tuveson showed in *Millennium and Utopia* that much Enlightenment thought married imagery of a clockwork universe (1972, 119–120) with the idea that science could uncover the secrets of a 'happy Fall' and thus help regain a lost paradise on earth (156–157). Adorno and Horkheimer famously showed the way that such high ideals were twisted into a machine-like transformation of the earth, the workers and even the individual's psyche into profitable units of trade by capitalist industries (2002 [1944]). It is in the context of these powerful ideas and such outcomes that we need to consider the modern desire to live in permanent light (Berry 2010).

But of course the process by which abstract ideals of light are concretised into physical living conditions designed to overcome the darkness of the lived moment occurs at mundane levels as well as in grand projects to dispense with the night. The everyday commodity represents this manifestation to Bloch, who was well aware of the myriad of synthetic analogues that could be offered to replace a lost sense of embodied transcendence or expansive, liberatory immanence in psychological or material terms. In regard to the 'inner' person, he noted that generally, people want to make a good impression and attempt to 'shine in front of others', the ego transforming itself into 'a saleable, even sparkling commodity' in an effort to be embraced (1995, 339). Unsurprisingly, Bloch sees this personal, inner realm of desired lustre presented back to us as a product to be purchased, whether we spy it from the comfort of our own home or out on the street: light 'calls' us in its guise as the commodity, 'dazzlingly illuminated behind glass, looking for customers' (342). The 'lizard-skin shoes trimmed with chamois leather' capture the attention of the woman, which inspires the man to look at her in turn, so that both are instantly enmeshed in a wish for more, which is ignited but rationed (34). The modern manifestations of these desires gleam across the face of the industrialised planet like a multi-faceted jewel of shopping malls and the commodities on display within them, but it is also manifest in the light that signifies the purported goodness of technological progress: of city buildings lit at night with the glow of prestige even when empty, of the street lights and factories that provide people with that sense of security and the promise of permanent abundance that is implicit in the capitalist paradigm.

The engine driving this cornucopia of materialistic efflourescence is industrialization fired by fossil fuels. Electric light was early recognised as the guiding icon of modern progress, dispelling the dark to symbolise humanity's newly won cultural power over the rest of nature, in terms both physical and metaphorical at once (Nye 1991). The association between the new magic of electric light and the utopian dreaming of modern industrial culture was made explicit during the glory years of the US Worlds Fairs. David Nye showed how early lighting displays went from being deeply impressive public spectacles to having their dazzle quickly neutralised in the eyes of a public hungry for more marvels. This forced exhibitors to increasingly heighten the spectacle of their annual displays as the unstoppable allure of electricity spread across the West during the early twentieth century (Nye 1991, 161–162, 368–371). The same effect soon after became apparent in terms of street lighting; from crowds flocking to the latest American city that had installed enough electric light to become known as a 'Great White Way' street, to a generally held expectation that urban activities should now extend into the night (29, 34, 56–57, 353). The ability to ignore agricultural rhythms and natural cycles of day and night has even become a useful way to define what a city is, according to Tuan (1978, 1–2).

This was the modern light that would finally free us from the cloak of darkness that limited night time activities to whatever could be carried out by feel, fire or candle light. Yet perhaps its terrible irony – the knowledge we now have that all this furious incandescence would eventually blanket the atmosphere with the greenhouse effect – was always in train. The human fascination with fire and its combination of stimulating and threatening aspects have been encapsulated in mythic literature since time immemorial. This classic narrative pattern, often recognised as a cautionary tale about our material and symbolic practices at once, is most recognizable to western audiences in the mythic tales of Prometheus, Icarus and Phaethon. It was of course modernised in Mary Wollstonecraft Shelley's *Frankenstein; or the Modern Prometheus* (1818). This version of the metaphor continues to remain relevant to the twenty-first century, as the crazed doctor sought to replace gods of both sky ('transcendental' creation as if from above and beyond, the abstracting dream of technoscience) and earth ('immanent' creation out of the womb). His scientific genius then replaces the masculinised and feminised realms of symbolic generative power – the heavens and the earth submit to the new bearers of cultural technology. If only we could steal fire from the heavens and earth concurrently, we could hope to 'pour a torrent of light into this dark world' (1992 [1818], 52).

Industrial technoscience disenchants the world, yet cloaks itself in a halo of magic in this act of transformation. As Wolfgang Schivelbusch noted, the disenchanted magic of electricity calls up and delivers the ancient promise of fire, transforming 'the flame's magical power of attraction' into a miraculously safe and instant form of light (1995, 179). While many no doubt feel that the sense of enchantment that accompanies firelight may have been removed by artificial illumination, commentators note that the subtle feeling of magic conveyed in the 'flick of a switch' has not been extinguished (Schivelbusch 1995; Brox 2011). The combination of light, magic and a feeling of mastery over the earth combine well with the fantasy that we could attain perpetual power. In the British Electrical Development Association's 1954 newsreel film *Out of the Dark: The Story of Lighting*, we learn that the emerging technology of fluorescent lamps promised to shed less heat than former bulbs. The dream that appeared as this process's logical conclusion was a combination of the best in nature and culture: it aimed to 'produce light without any heat at all', as does the firefly. The magic of electric light thus helps convince us we can utilise the natural world to become free of it, preferably avoiding the cost (in this case the combustion of dangerous levels of greenhouse gases). Bloch would be delighted, having suggested in *The Principle of Hope* that: 'The two most favourite general wishes of mankind are to stay young and live long. And a third is precisely to achieve both, not in a painful roundabout way but with a fairy-tale quality of surprise' (1995, 455).

But even if we could create light without heat, or any other kind of waste, we still require some kind of fuel (at least for the foreseeable future). Not only that, recent history suggests that even a perfectly sustainable way of producing light would be accompanied by a drive towards more consumption in general. The mystique and prestige of light, which conveys an aura of abundance in physical and symbolic terms concurrently, has survived the transition of western European life from the hearthfire of a simple cottage in the woods to the electrically lit suburban apartment in a metropolis. Although the flickering of flame was made redundant by the relatively recent invention of clear, instant electric light, the draw of a home filled with luminescence could be even stronger in today's world than it was when procuring and maintaining a flickering hearthfire actually presented a real (and messy) challenge (Brox 2011).

Just as associating living 'in the light' with an existence in utopian paradise is common to religious and philosophical discourse, the suggestion is also found in the most humble narratives of home – the warmth of the hearth, the cordiality of kindred spirits in a domestic setting. The light of home represents something attractive to both inner and outer nature; it grants a feeling of comfort and security that can soothe us both spiritually and physically after a day 'out in the world'. The age-old wisdom that fire is both boon and threat emanates from humanity's immediate experiences with it – it endangers us as the escaped flame from a hearthfire that destroys home and precious belongings, or as the forest fire that engulfs a valley yesterday filled with life. But in the twenty-first century era of ecological destruction and climate change, this threat is extended beyond the immediacy of the flame and individual people or places. Today, the way we kindle fire in accord with an innate desire for an abode filled with light threatens the very viability of life on our wider home – the planet.

In this advertisement from European telecommunications company UPC (which was ubiquitous during my time in Ireland in 2012), an electric blue river of light flows in the front door, offering the magic of instantaneous access to signals from (and to) anywhere around the world. The electronic age, while still (in fact, increasingly) fired by fossil fuel power[1], transports its consumer to a world where time and space have been overcome, condensed into the immediacy and completion of material being uploaded from anywhere that is connected (or filmed by those with access to the new 'sky gods' of the 'tower cult'

The utopian home 2: fibre cables, as light, representing faster information and communication. UPC, Ireland, 2012.

– the combination of camera and satellite; Eisenberg 1998, 434). In the post-Enlightenment, truly 'modern' spirit of desacralised embodiment for sale, the way this vision is materialised places the commodity in the most eye-catching light. It represents transcendence without delay and materialism without limit (for those capable of and prepared to exchange their labour power for currency). Thus the physical distance between the consumer and the raw materials that went into the commodity's production (whether it be a physical product or an electronically delivered image) remains a matter of some import. Marx was right to see that this was fundamental to the mystique of the commodity as well as the alienation of the workers; the element of magic with which the commodity is embossed requires a rejection of the actual conditions under which it appeared (Marx 1970 [1867], 75–80). Naturally, as Bloch saw 80 years ago, the 'big drum' for this is advertising, which 'makes magic out of the commodity, even the most incidental commodity, a magic in which each and every thing will be solved if only we buy it' (1995, 343–344). But as Bloch knew and Levitas reminds us, advertising works because it keys 'into utopian images which are already present among the audience, reflecting their desires, their lack' (1990, 189).

Modern light, as the eternally beneficent symbol of creation, life, the good, truth, justice and forthrightness, is materially manifest in the street light, the well-lit corporate building, the neon billboard, in the television and computer screen and in the glow of the mobile phone. The commodity is now image, accessibility and service provision as well as product and the city has become, according to Jameson 'a fundamental form of the Utopian image' (2005, 4). Urbanity now provides a space of corporeal transcendence, wherein 'even the most subordinate and shamefaced products' offer 'muted promises of a transfigured body' with 'overtones of immortality' (6). But of course this is a gigantic engine room of deceit, a Platonic cave within which modern urbanites dwell in increasing darkness and ignorance of a greater, truer light outside the illusion. Blumenberg showed the way that artificial light has resulted from a long process of technological manipulation, which has resulted in a world of optical prefabrications and fixations of the gaze that approach that of Plato's cave (1993, 54). We are also reminded that Adorno and Horkheimer critiqued the 'Culture Industry' as a mechanical monster, one-dimensionally unanimous in its subjugation of the individual in the same kind of master/slave relationship as settlement civilisation assumes over the rest of nature (2002, 94; "Enlightenment as Mass Deception" was the subtitle to this chapter).

The brightly lit mythos of consumption cannot admit that it is limited by the earth's capacity to continue to fuel its fires, for this would undermine the very utopian telos it represents – that we can be free of the limits of the body while still inhabiting it. Nowhere is this urban light and the commodity it highlights placed on show better than in modern shopping malls. Carolyn Merchant defined their aura of perfectly lit abundance as an attempt to re-attain a state of harmony with nature in a new, desacralised earthly garden of plenty (2003, 167–168). She saw this as a repetition of the mythic model of heroic evolution, or the 'upward progress of humankind from darkest wilderness to enlightened mind' (178–179). But of course the drive to recreate a Paradise Lost with the city comes with ecological damage done in real, physical terms to the earth. A brilliant fictional account of the way the privileged get to inhabit the redesigned Eden, while the earth and many of its 'others' suffer, is provided by Kenneth Burke to highlight this unfortunate truism. Burke satirises hyper-technological utopianism in the figure of Helhaven, a combination of the Garden of Eden and the Tower of Babylon on the moon, with artificial vistas programmed into the view from its Luna-Hilton Hotel windows (2000, 96–103). Like Orwell's *1984*, though, the material for this satire already existed at the time of writing; the tale is, then, a symbolic representation of the twentieth century's direct or indirect profit derived from 'the polluting of some area' or another (101). Hence the Chosen, indicated in the New Jerusalem of Revelations or as the inhabitants of Dante's Paradiso, are those lucky enough to have been responsible for the despoiled earth they've left behind, where the poor remaining are not the Sinners but simply the unlucky (102). Havoc is wreaked on the environment of the poor in order to fund the 'bubble' of idealised, elitist Arcadia for those who can afford it (Coupe 2000, 65). Val Plumwood agrees that the ecological cost of materialistic consumption tends to be passed on to the less privileged, often in far distant 'shadow places' (Plumwood 2008). Heise likewise compares the utopian promise of technology in 'the global North' with the dystopian environmental 'scenarios of pollution, deprivation and oppression' in other parts of the world (Heise 2012, 3).

This equation, whereby we generally gain from ignorance of the true conditions that underwrite our material abundance, is in direct contrast to the kind of collective awareness of production and consumption processes that Bloch would require if we were to inhabit a true 'homeland'. Because this hopeful space requires that the Alpha and Omega of its conditions of possibility become apparent to each other at the same time, it cannot be encapsulated in archaic or archetypal imagery such as Plato's anamnesis, to a return to an 'immemorial dimension of a perfection in order to fill its Optimum with it' (1995, 305). The true homeland towards which we must continuously move is 'the *purely utopian archetype, which lives in the evidence of nearness, to that of the still unknown, all-surpassing summum bonum'* (305). This highest good is 'the most felicitous astonish-ment' and its possession transforms the moment 'into its completely resolved That' (305).

Critical and creative ecotopias and ecodystopias

A mature attempt at thinking through the possibilities inherent in the utopian imaginary in relation to the ecological crisis entails at least two considerations. It must avoid the perils of the pre-critical phase (unhealthy fixations on overt romanticism, pastoral longing and nostalgia, infantile fantasy and so on), and it must stand for hopes of ecological regenera-tion, biodiversity, and an atmosphere not made ruinous by carbon pollution. As Jameson noted, a new utopian content will doubtless arise from the life world of those living close to the land, the plebeian world 'of growth and nature, cultivation and the seasons, the earth and the generations' (2005, 85). But as we are all aware, dystopian fiction and film have for some time far outweighed hopeful imaginings of the future. James Cameron's film *Avatar* became the highest grossing film of all time after its release in 2009 and it conveyed a vision of humanity as a greedy colonising force, led by a partnership of military and corporate interests who were ruthlessly willing to destroy an ecology and indigenous people for profit. One need not have stared into the mirror of modernity for too long to see ourselves – the constituents of technologically advanced societies – cast as the villains here, even as many viewers sympathise with the native Na'vi inhabitants of Pandora. The beauty of the moon's habitat is clearly designed to inspire such admiration, as is the soulful nature – dare we use the word honour? – of the Na'vi people and their relationship to the other creatures and life forms of the land. Commentator Bron Taylor investigates the biocentrism and the controversies around the 'dark green' politics of the film in *Avatar and Nature Spirituality* (2013).

More recently, the best-selling *Hunger Games* trilogy of Suzanne Collins is being released in film format and deals explicitly with the way 'shadow places' are harvested to fuel the satisfaction of a more privileged population's appetites. This is an important development in the field of the critical ecodystopia because it places Katniss Everdeen's 'District 12' as coal mine to the distant Capitol's splendour (coal being the industrial age's real version of Pandora's fictional mineral 'Unobtanium'). The 'city utopia', as Sargent calls it, requires fuel to fund its abundance, which in turn offers the inhabitants of the Capitol more modern 'human contrivance[s]' and thus offers a cultural path towards the realization of the more ancient 'body utopias, or utopias of sensual gratification' (Sargent 2007, 301–302). As Sargent points out, this enables humanity to replace the beneficence of nature or the gods and thereby assert their independence. It also, in the *Hunger Games*, allows the Capitol's inhabitants to follow outrageous fashions and indulge in gossip dismissed as desperately inane by Katniss, used as she is to the harsh life and coal dust of District 12. Add Marx and bingo – you can no longer imagine beautiful urban places free of trade relationships, unlike Ebenezer Howard's utopian novels *Tomorrow* (1898) and *Garden Cities of Tomorrow* (1902). These fictional societies included newly orga-nised social functions, but failed to explain the power sources that kept them functioning (Bloch 1995 [1959], 612). Howard's utopias explicitly stated that a healthy biosphere was required for healthy human bodies and minds; both required air, light and sunshine to 'cure the defects at home' (612). This makes them ecotopian, but not critically so. By contrast, the *Hunger Games* depicts a (North American) world where fuel is paramount and the relationships between the elite (dwellers of Panem's Capitol) and the people of the land (in this case District 12) are manifestly unequal. As in *Avatar*, citizens of technolo-gically advanced urban places are asked to question their privilege and the relationships to land and others that fund it.

If *Avatar* and the *Hunger Games* work as warnings of what can be done to the earth and the people of a particular place when profit and the politics of mastery or domination hold sway, then Ursula K Le Guin's *Always Coming Home* (1985) represents a different set of ecotopian/ecodystopian possibilities. The novel features a people called the Kesh, who inhabit a valley in what was once California, in a future so post-apocalyptic that much healing has been done (although painful memories arise with the 'Sevai', a fatal nerve condition 'related to residual ancient industrial toxins in soil and water' (304). The sea is raised significantly and most folk live in a way that reflects Le Guin's indigenous and Taoist leanings. This does not render them immune from conflict, however. Realistic appraisals of human weaknesses include the Kesh prejudice against the nomadic tribes who passed through their lands, while the village lifestyle also features small-minded superstitions and petty jealousies as well as rituals to alleviate such facts of life. Yet the Kesh society enjoys a kind of socialism according to which a materially poor family can take more than they provide from the collective stores. Totemic animals and plants are used to divide and join together different groups of people and other animals (who are thought of as kin) in a taxonomic system that is thought through to the most complete and minute detail. The Kesh 'Dream Time' teaches that people are animals and animals (as well as plants and stones and water) are people, not in an anthropomorphic sense but as kin related through shared participation in life. Yet, as Lisa Garforth notes, 'wild nature' is retained as paradoxically and fundamentally other; the Kesh answer the 'paradox of radical ecological philosophy' by identifying with it while it remains separate insofar as it is accorded intrinsic value (2005, 417).

The multilayered word 'heya' works to celebrate the mystery of the 'hinge' between worlds and seasonal rituals continuously reinvigorate Kesh culture around times of death and rebirth, increase and letting go, marriage and moving on and so forth. The Kesh find ways to live with grief and loss, tragedy and the scatalogical, recognising that a desire for purity runs against the realities of life on earth (188, 192, 201, 219, 312). Their ecospirituality is loyal to the earth yet open to the mysteries of incarnation and human relationships with the 'otherworld'. Flicker, a visionary character of the Serpentine of Telina-na, explores this paradox, which could be considered vital to a fulsome ecotopian imaginary. The novel features a sense of panpsychism, or animism, which infiltrates all time or space but finds special composition in sacred sites. In the short play 'The Plumed Water', fire and water meet in the darkness beneath a celebrated geyser, in a conflict that creates light (221–226). As the spirit ancestor Puma breathes, the two elements meet and leap towards the sky, the geyser shining with 'that which falls from the sunlight as it rises from the dark' (225). It is both here and not here: 'As it lives, it dies' (226). Resisting simplistic literalisations of the mystic qualities of light, Le Guin has the Kesh recognise the positive value of darkness. By association, they live with a realistic awareness of the earth's limits when it comes to supplying human wants and needs.

Their post-apocalyptic way of being in place represents a path that profits from listening to the earth and working *with* it rather than standing *over*, or 'outside' of it. This quality is perhaps the single greatest gift of the ecotopian imaginary to the post-modern urban consumer. It can be compared in richness to indigenous epistemologies, which likewise ritualise transformations from one state or life stage to another, such that deeper satisfaction can be experienced in embodied experiences of sufficiency rather than endless yearning after materialistic abundance. This differentiation was made clear by Marius de Geus in *Ecological Utopias: Envisioning the Sustainable Society*, where he pointed out that utopia could be envisioned as a much more sustainable possibility if we learnt to live with less rather than desiring more. This is often and profitably imagined as the difference between William Morris's *News from Nowhere* (1890) and Edward Bellamy's *Looking Backward* (1888). The relationship to the 'education of desire' and philosophies such as E. F. Schumacher's *Small is Beautiful* are immediately apparent. The Kesh appreciate the mystery of life to a level at which acquisition becomes an activity of lesser status than altruism (in a real, everyday sense rather than as an ideal). They are grounded in a realization that the light we wish to inhabit is within, not in terms of some ancient religious ideal but as an ongoing flickering of the mystery of consciously self-aware life in an embodied form.

Reading *Always Coming Home* reminded me of the Australian Aboriginal 'songlines' – or 'country lines' as they are now known – recorded by anthropologist John Bradley.[2] For the Yanyuwa of far north-eastern Arnhem Land, the earth and sea are relatives, and all creatures sing together. The Yanyuwa elders asked Bradley to record some of their country lines, or *kujika*, in an effort to get the youth of new generations interested in their culture – the lore of which has been dying with their language. Ironically, the children are more inclined to take notice of their millennia old culture now that it is on screen in high definition CGI (computer-generated imagery) animation. Yanyuwa *kujika* are in and of the land and the sea. They represent complex casts of characters and the relationships amongst them, because Yanyuwa traditions recognise in relations between humans and their country a consistent negotiation between kin. This necessitates listening in to an enlivened spiritual cosmos filled with animate, sentient life forms (Bradley 2001, 297–298). It is believed that *kujika* are received, not constructed; they have their own agency, even without being sung, and are not mere expressions of the function of human survival. They travel through the deep sea and inhospitable mangroves, not with human kin in mind but as the song of the country (Bradley 2008a, 28–29). They include conflicts that are not always neatly or non-violently resolved because sometimes particular interests need to be forced 'into line' (personal communication).

In contrast to early or unsophisticated misunderstandings of ethnographic data, Australian Aboriginal country lines are not mere propitiations of the spirits or animist rituals to appease mysterious or threatening powers. Neither are they just about adhering to the lore of the past, although they are historically conservative. *Kujika* are ever-present in country; while placed there by the spirit ancestors (or Dreamings) in the deep past, they are not an endless repetition of cycles always looking back to a pristine age now lost, but *wandayarra a-yabala:* following a road both pre-existing and alive in the moment (28). Humans sing this song as a ritual of re-creation that enables them to experience the world anew. Bradley coined the term 'supervitality' to help explain this cultural process in a way that neatly supersedes the dualistic western division between realms of the sacred and profane (Bradley 2008b, 635). The performance of *kujika*, according to this model, takes the vitality innate in the land and concentrates it into a kind of supervitality. In the animated *kujika Mannannkurra*, or *Tiger Shark Dreaming*, a series of dancing lights accompany the song as it traverses land and water (in this case a coastal river; Yanyuwa Families et al. 2008). These *wirrimalaru* represent a permanent and abiding force, created by the continuing presence of the Dreaming ancestors. Their brilliance signifies physical, emotional, and spiritual health in the country. Light arises out of the earth and dances within it and amongst its other forms of life.

Given that there are no immediate provisions for dealing with the problems of global modernity built into this formula of indigenous wisdom, it would be the job of an ecotopian imaginary to translate such understanding into terms suitable to the worlds of the globalised urban consumer. This would require that the links between urban consumption and rural production (for example), which predicate the real costs of abundance, be made explicit. This in turn returns us to the problematic of the NASA imagery and the 'Dreaming' of global civilisation and its drive towards a utopia of abundance (or perpetual consumption). Creatively inspired critical ecotopians are required to think through the structure of the city and the way it works, what the grids of electricity and data and buildings say about the way we live, the significance of linear streets and malls and cars in terms of latent utopian desires, and what alternatives may be on offer. The skill of close listening to the earth, in the sense of local place and as a global set of relationships between people and places, would also grant us greater appreciation of the damaging aspects of the excessive light pollution and media white noise that afflicts the urban megalopolis. All of this requires, for the critical ecotopian, keeping in mind that we are immersed in and dependent upon the rest of nature, yet at some level we seem impelled to define ourselves as separate from it; and this desire must be appreciated alongside the alienation it entails. In conclusion, when the critical ecotopia responds to a global perspective on multiple levels, combined with the patient observation of the small, local and gentle as well, it resonates as a mature response to the needs of the times and to the tradition of the country lines. From deep within such an ecotopian worldview, we may see fit to release ourselves as a people from the fuel fetish/light worship/power dream of perpetual abundance that continues to drive the postmodern, global urban marketplace.

Notes

1. The OECD, for instance, envisages that greenhouse gas emissions from such sources will increase by another 50% by 2050; Marchal et al. (2011, 5).
2. 'Country' is a complex term in Australian Aboriginal epistemologies, denoting the kind of living concept of land, sea and creatures Bradley describes here. It also involves relationships of responsibility laid down across the living world of this 'eco-cosmology' by the creative, ancestral dreamings (Bird Rose 2005).

References

Adorno, T. W., and M. Horkheimer. 2002 [1944]. *Dialectic of Enlightenment: Philosophical Fragments*. Stanford, CA: Stanford University Press.

Bellamy, E. 1888. *Looking Backward 2000 – 1887*. Accessed April 3, 2012. http://www.gutenberg. org/ebooks/624

Berry, G. 2010. "Under the Dominion of Light: An Ecocritical Mythography." PhD thesis. Accessed November 20, 2012. http://arrow.monash.edu.au/hdl/1959.1/166997

Berry, G. 2012. "Urban Light and the Ecological Crisis: Planetary Imagery and the Night." Proceedings of the Reading Nature; Cultural Perspectives in Environmental Imagery conference, Universidad Complutense de Madrid, published by Friends of Thoreau – Franklin Institute, 26–34.

Berry, G. 2013. "The Symbolic Quest Behind Today's Cities of Light – and Its Unintended Ecological Consequences." *Journal for the Study of Religion, Nature and Culture* 7 (1): 7–26. doi:10.1558/jsrnc.v7i1.7.

Bird Rose, D. 2005. "Pattern, Connection, Desire: In Honour of Gregory Bateson." *Australian Humanities Review*. Accessed September 17, 2013. http://www.australianhumanitiesreview.org/ archive/Issue-June-2005/rose.html

Bloch, E. 1995 [1959]. *The Principle of Hope*. Translated by Neville Plaice, Stephen Plaice and Paul Knight. Cambridge, MA: MIT Press.

Blumenberg, H. 1993. "Light as a Metaphor for Truth." In *Modernity and the Hegemony of Vision*, edited by D. Levin, 30–62. Berkeley: University of California Press.

Bradley, J. 2001. "Landscapes of the Mind, Landscapes of the Spirit: Negotiating a Sentient Landscape." In *Working on Country: Contemporary Indigenous Management of Australia's Lands*, edited by R. Baker, J. Davies, and E. Young. Melbourne: Oxford University Press.

Bradley, J. 2008a. "Singing Through the Sea: Song, Sea and Emotion." In *Deep Blue: Critical Reflections on Nature, Religion and Water*, edited by S. Shaw, and A. Francis. London: Equinox.

Bradley, J. 2008b. "When a Stone Tool Is a Dingo: Country and Relatedness in Australian Aboriginal Notions of Landscape." In *Handbook of Landscape Archaeology*, edited by B. David, and J. Thomas. Walnut Creek, CA: Left Coast.

Brox, J. 2011. *Brilliant; the Evolution of Artificial Light*. New York: Souvenir Press.

Burke, K. 2000. "Hyper-Technologism, Pollution and Satire." In *The Green Studies Reader*, edited by L. Coupe, 96–103. London: Routledge.

Coupe, L., ed. 2000. *The Green Studies Reader: From Romanticism to Ecocriticism*. London: Routledge.

de Geus, M. 1999. *Ecological Utopias: Envisioning the Sustainable Society*. Utrecht: International Book.

Durst, J. (Director). 1954. "Out of the Dark; the Story of Lighting." Film produced for the British Electrical Development Association by Merton Park Studios in association with the Film Producers Guild (viewed on site at the Australian Centre for the Moving Image with thanks to their archival materials collection).

Eisenberg, E. 1998. *The Ecology of Eden*. New York: Alfred A. Knopf.

Frye, N. 2004. *Biblical and Classical Myths: The Mythological Framework of Western Culture* (with Jay MacPherson). Toronto: University of Toronto Press.

Garforth, L. 2005. "Green Utopias: Beyond Apocalypse, Progress, and Pastoral." *Utopian Studies* 16 (3): 393–427.

Geoghegan, V. 1997. "Remembering the Future." In *Not Yet; Reconsidering Ernst Bloch*, edited by J. O. Daniel, and T. Moylan, 15–32. London: Verso.

Heise, U. 2012. "The Invention of Eco-Futures." *Ecozon* 3 (2): 1–10.

Jameson, F. 2005. *Archaeologies of the Future; the Desire Called Utopia and Other Science Fictions*. London: Verso.

Le Guin, U. K. 2001 [1985]. *Always Coming Home*. Berkeley: University of California Press.

Levitas, R. 1990. *The Concept of Utopia*. Hertfordshire: Syracuse UP.

Marchal, V., R. Dellink, D. van Vuuren, C. Clapp, J. Château, E. Lanzi, B. Magné, and J. van Vliet. 2011. "OECD Environmental Outlook to 2050: Climate Change Chapter." http://www.oecd.org/dataoecd/32/53/49082173.pdf.

Marx, K. 1970 [1867]. *Capital: A Critique of Political Economy*. Vol. 1 of *A Critical Analysis of Capitalist Production*. London: Lawrence & Wishart.

Merchant, C. 2003. *Reinventing Eden: The Fate of Nature in Western Culture*. New York: Routledge.

Morris, W. 1890. *News from Nowhere*. Accessed March 30, 2012. http://www.gutenberg.org/ebooks/3261

Moylan, T. 1997. "Bloch Against Bloch." In *Not Yet; Reconsidering Ernst Bloch*, edited by J. O. Daniel, and T. Moylan, 96–121. London: Verso.

Nye, D. E. 1991. *Electrifying America*. Cambridge, MA: MIT Press.

Plumwood, V. 2008. "Shadow Places and the Politics of Dwelling." *Australian Humanities Review*, no. 44. Accessed May 28, 2011. http://epress.anu.edu.au/ahr/044/pdf/eco02.pdf

Sargent, L. T. 2007. "Choosing Utopia: Utopianism as an Essential Element in Political Thought and Action." In *Utopia Method Vision; the Use Value of Social Dreaming*, edited by T. Moylan, and R. Baccolini, 301–317. Oxford: Peter Lang.

Schivelbusch, W. 1995. *Disenchanted Night; the Industrialization of Light in the Nineteenth Century*. Translated by Angela Davies. Berkeley: University of California Press.

Schumacher, E. F. 1973. *Small Is Beautiful*. Michigan: Blond and Briggs.

Shelley, M. W. 1992 [1818]. *Frankenstein*. London: Penguin Books.

Taylor, B. 2013. *Avatar and Nature Spirituality*. Waterloo, ON: Wilfrid Laurier University Press.

Tuan, Y. -F. 1978. "The City: Its Distance from Nature." *Geographical Review* 68 (1): 1–12. doi:10.2307/213507.

Tuveson, E. 1972. *Millennium and Utopia: A Study in the Background of the Idea of Progress*. Gloucester, MA: Peter Smith.

Yanyuwa Families, J. Bradley, B. McKee, C. Ung, and A. Kearney. 2008. *Manankurra*, DVD, directed by J. Bradley. Melbourne: Monash University. Accessed August 21, 2013. http://www.infotech.monash.edu.au/research/projects/independent/countrylines-archive/animations/song-tiger-shark-manankurra-yanyuwa/

Index